THE
SCIENCE
OF
MILITARY POSTS,
FOR THE USE OF
REGIMENTAL OFFICERS,
WHO FREQUENTLY COMMAND
DETACHED PARTIES,
IN WHICH IS SHEWN
The manner of ATTACKING and DEFENDING
POSTS.

M. LA COINTE,
of the ROYAL ACADEMY at NISMES
1761

The Naval & Military Press Ltd

published in association with

FIREPOWER
The Royal Artillery Museum
Woolwich

Published by
The Naval & Military Press Ltd
Unit 10 Ridgewood Industrial Park,
Uckfield, East Sussex,
TN22 5QE England
Tel: +44 (0) 1825 749494
Fax: +44 (0) 1825 765701
www.naval-military-press.com

in association with

FIREPOWER
The Royal Artillery Museum, Woolwich
www.firepower.org.uk

The Naval & Military Press

MILITARY HISTORY AT YOUR FINGERTIPS

... a unique and expanding series of reference works

Working in collaboration with the foremost regiments and institutions, as well as acknowledged experts in their field, N&MP have assembled a formidable array of titles including technologically advanced CD-ROMs and facsimile reprints of impossible-to-find rarities.

In reprinting in facsimile from the original, any imperfections are inevitably reproduced and the quality may fall short of modern type and cartographic standards.

THE FRENCH AUTHOR dedicates his work to the prince of Conti, having, as he says, acquired his knowledge under his highnefs in Piemont and Flanders.

❀❖❀❖❀❖❀❖❀❖❀❖❀❖❀❖❀❖❀❖❀❖❀❖❀

The following pieces, printed before the original, we think proper to give with the tranflation, that the reader may fee what value was fet upon it in France.

A letter from marfhal count de Lautrec to M. la Cointe, captain of cavalry.

SIR,   Paris, Sept. 25, 1758.

I Return you your little treatife in manufcript on the defence of military pofts. I read it with attention, and I prefume, if you make it public, it will be favourably received; for what you relate on this fubject is very inftructive, and young officers may from thence learn principles proper to direct them how beft to preferve pofts that are entrufted to them; which, by their fitua-
                                                    tion

tion, are often of infinite confequence, and contribute much to the fafety of the camp, as well as to the army on its march; befides feveral other occafions wherein they may be of great utility.

This, Sir, is what I think of the little work on which you have confulted me; your application, and the zeal you fhew for the king's fervice, are equally commendable. I wifh that the court, knowing your merit and your talents, may not let them lie idle. You need not doubt, that for my own part, I fhall take all opportunities to recommend them; affuring yourfelf, Sir, that no body can be more difpofed than I am to oblige you on all occafions.

Le Marechal de LAUTREC.

The extract from the regifter of the royal academy at Nifmes,

Contains no more than an acknowledgment of M. la Cointe's being of their academy, and allowing him, after fubmitting his work to be examined by a committee of their *members*, to publifh it with his character of Academift in the title page.

Appro-

Approbation of Messrs. the committee of the academy of Nismes.

WE the underfigned, commiffaries named by the royal academy at Nifmes, to examine a work of M. la Cointe, intitled, *the fcience of military pofts*, &c. certify, that we have read this work with attention. It has appeared to us the more ufeful and inftructive for young officers, becaufe no one has hitherto methodized that branch of the art of war, which is the fubject of this book, into principles that may direct their practice; which is a matter of great importance, from the connection between it and the grandeft operations of armies.

Thefe motives engage us, in the name of the royal academy of Nifmes, which has impowered us fo to do, to permit M. la Cointe to take in his work the title of Academift of Nifmes. In witnefs whereof we have given the prefent cer-certificate. Paris, Feb. 20, 1759.

Le marquis D'AUBAIS MENARD.

Appro-

Approbation of M. Belidor, brigadier
of the king's armies, cenfor royal for
the artillery and engineers of the
academy of Berlin.

I Have read, by order of my lord
the chancellor, a manufcript, intitled, *the fcience of military pofts*, by M.
la Cointe, formerly lieutenant of infantry, fince captain of cavalry. This work,
wrote with a great deal of care, method and erudition, comprehends the
beft maxims that can be given on the
manner of fortifying and defending the
advanced pofts of an army; it will
therefore be of great utility to young
officers; as I know of nothing more
inftructing, or more proper to excite
emulation. Given at Paris this 19th
of Feb. 1759.

BELIDOR.

Dedi-

DEDICATION.

To the fubaltern officers in the British army.

GENTLEMEN,

A Severe illnefs, in confequence of fome hard campaigns, confined me for feveral months to my chamber; as much of this time as my diforder would permit, I began to employ in ftudying fuch authors as might be of ufe to me in my profeffion ; among which, the little piece which I now prefent you with, feemed fo well calculated to be the pocket companion of a young foldier, that I thought, though I was unable to do my duty in the field, I might in the mean time do an acceptable fervice to my brother officers, by recommending Mr. le Cointe's leffons to them in our own language.

I am, GENTLEMEN,

Your fincere well wifher,

and moft humble fervant,

The TRANSLATOR.

Tranſlator's PREFACE.

THE tranſlator had rather be blamed for a bad ſtile, or for copying the French idioms, than run the riſque of changing the ſenſe of the original by too much poliſhing: a ſentence brought from one language to another, when relative to ſcience, ſhould be changed as little as poſſible, for fear of inverting the ſenſe ; however, in ſome places, eſpecially in the geometrical part, he was obliged to help his author; but for the reſt, as his intention was to give a faithful tranſlation, whatever thoughts of his own occurred, he has given them by way of notes at the bottom of the page

As for the merit of the work itſelf, ſince the foregoing approbations certify it to be an original work in France, we apprehend it will be to the full as new and as uſeful to young military men in Great Britain, where the theory of the art of war ought to be the more carefully cultivated, as the happineſs of our ſituation, and mode of government, give us, in compariſon with our ambitious neighbours, but few opportunities of practiſing it. IN-

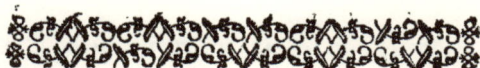

CONTENTS.

CHAP. I.

CHAP.

The CONTENTS.

CHAP.

The CONTENTS.

CHAP IX

CHAP X.

The

The CONTENTS.

INTRODUCTION.

THE ambition which animates our young military people is commendable, and becomes every day a gain to the ftate. Excited by examples, which they have before their eyes, to make their way to honours and favour, and convinced, that capacity and talents entitle them to fucceed, they are lefs occupied with amufements that wafte their time, and have more application

This idea has produced a happy change, and we now fee more emulation and zeal than ever there was before ; almoft all officers ftudy, almoft all officers draw ; and excepting a fmall number, who look upon the fervice as

B   a life

a life of independence, wherein they
may take the liberty to neglect all the
sciences; there are few who do not feel
how advantageous it is to put themselves in the way of being known.
The progress of reason has had an in-
fluence on all the arts; and to appear
on the parade, and be master of the
manual exercise, is no longer looked
upon as the only merit necessary in a
regimental officer, because we see these
duties performed by soldiers ever so little disciplined.

A man that would advance himself,
must study every branch that belongs
to the art of war.

As the end that every one proposes,
when they enter into a profession, is to
advance themselves therein; the whole
care of a young officer should be to in-
struct himself in that of war.

As I was designed for that profession
from my earliest youth, I learned be-
times, and have studied ever since, that
part of mathematics which is most important for a young officer to know.

This

This application, and twelve years experience in the foot fervice, having caufed me to make reflections on the fortifications of pofts, to which regimental officers may be detached, I have given them to the world, perfuaded, that they may be ufeful and advantageous to the King's fervice.

Another reafon determined me thereto: having feen, in the different detachments that I made during the laft war in Piedmont and Italy, how much a young officer, who has no idea of fortifications, is embarraffed, when he is ordered to intrench himfelf; I thought a book, containing principles by the help of which fuch works may be eafily performed, and which would give, at the fame time, the methods of defending and attacking fuch pofts, would be a very great help to them.

No author, that I know of, has hitherto laid down thefe principles, fo as to render them immediately ufeful to young officers; the moft of them have feemed as if they intended only to give leffons to Generals, by writing profound

treatifes

treatifes on the grand operations of an army, and as if they difdained to expatiate on fuch as they imagined were lefs important.

Chevalier Follard, and Chevalier de Clairac, are the only writers on the attack and defence of pofts of this nature ; but the former, whom we may confider as the reftorer of the true principle of war, has touched them but lightly ; and the rules, given by the latter, are fo connected with the greater works made in the field, fuch as intrenching of armies, lines of communication, and trenches, that they can be of little fervice to private or regimental officers.

The authors who have written after them, have gone no deeper into this part of the fcience, becaufe they did no more than either to copy or abridge the others, without ever entering into thefe particulars that the fubject is capable of. However, the fcience of pofts was always an object effentially neceffary to the *greateft captains.*

" It

" It is," fays the commentator on B I.
Polybius, " one of the principal qua- Ch. 14.
" lifications requifite in the commander
" of an army, and perhaps the leaft
" known."

I will add, that it is by the help of
this fcience only, that an army can en-
camp with fafety, that it may reft from
fatigues, and fcreen itfelf from the con-
tinual inquietudes that the enemy's
parties might give it.

It is now no longer a doubt, that
war, like other arts, is to be ftudied,
both in the clofet and by exercife ; a
thoufand examples have proved that an
officer, who applies himfelf both thefe
ways, has an infinite advantage over
another who goes on in the vulgar tract,
and learns only by rote.

" It is fine talk to fay to an officer,
" be firm and courageous, never re-
" treat, conquer or die ; thefe maxims
" and rules, fays M. de Botie in his
" treatife of Military Studies, make
" no impreffion on the heart of a man,
" but in proportion as his mind is en-
" lightened, by knowing the methods
" of

" of conquering, or blind to the dan-
" ger of being overcome."

In truth, it is ftudy that opens our
underftanding, and excites our application; it is by that we fupply our want
of experience, by that we acquire thofe
qualities which form great officers, and
by that we open to ourfelves the way to
renown.

In the general operations of war,
fuch as fieges and battles, the glory
afpired after by all military men, is re-
ferved for the fuperior officers only, becaufe in thefe great actions every thing
is attributed to them, and put to their
account. Therefore it is only when a
private officer, having the chief command of a party, can make a gallant
defence, or can execute an enterprize
to be talked of, that he may thereby
be the inftrument of his own glory,
may merit the commendations of the
army, and the favour of the court.

What fatisfaction muft a young foldier feel, when by various devices he
fo oppofes his enemy, that he fecures
himfelf from furprizes, refifts his attacks,

tacks, difconcerts his projects, and makes him abandon his enterprize ! Comparable then to the greateft captains, he has a fhare in the profperity of the arms of his fovereign, and in the defence of his country, and merits fo much the more, as one of our mafters in the fcience of war, afferts, " that the glory acquired in the de- " fence of a weak poft, is infinitely " above what may be gained in the " moft important fortreffes of the " ftate."

Follard, Vol. 5.

The means of acquiring this glory, never depends fingly or merely on the greatnefs of one's courage, which is ufeful only in the execution ; but on the combination of the talents that are neceffary both to contrive and execute a project with fuccefs.

That bravery which elevates us above all dangers, is not fufficient ; it may even turn to our damage, unlefs a wife and enlightened conduct reftrains us from rufhing thoughtlefs and rafhly into action.

There

There will be no detail in this treatife, of what belongs to thofe pofts, to which *general* officers are commonly detached; nor rules for the conftruction of lines to eftablifh a communication, and enfure the fafety of an army.

As I write only for *private* or *regimental* officers, I fhall treat only of fuch pofts, as they may be detached to, with thirty, fifty, or one hundred men; and will lay down, 1ft, Some general notions that they fhould have of geometry, to be able to trace out entrenchments. 2dly, The different works for fortifying pofts 3dly, How to augment their force on all occafions. 4thly, How an officer fhould be prepared to go on a detachment. 5thly, How he fhould march towards a poft that he is detached to. 6thly, How he may eftablifh himfelf therein. 7thly, What precautions he fhould take to prevent his being furprized therein. 8thly, What difpofitions he fhould make to maintain himfelf there with vigour. 9thly, The manner of defending pofts. 10thly, and laftly, How to to attack them either

ther

ther by open force, or to carry them off
by ftratagems.

Such is the plan of this work, from
whence, I do imagine, that inftruétions
may be drawn, to fortify, defend, or
attack, even the moft confiderable pofts,
as well as the fmalleft; the rules in
thefe refpeéts being the fame, and dif-
fering only in the fize of the works,
which muft be proportioned to the number of men, that the party confifts of.

Whatever fimilitude there feems to
be between the fervice in time of peace,
and the fervice in time of war; I will
venture to fay, that they will fcarce bear
a comparifon, and that they fhould be
rather confidered as two different pro-
feffions.

In garrifons the fervice is extremely
fimple, there being no more requifite,
than to know how to obey. In camp,
it is conneéted with a thoufand accidents, that require an officer to have
fkill to command well, and aét a determined part. Any one will be convinced
of this difference, when they have feen
my particular opinions, fupported by

C   general

general examples. The relation of facts
being of all methods of writing the
most useful, most instructing, and most
amusing; I will quote both good and
bad, analogous to each article: the latter, that we may learn wisdom at the
expence of those who have lived before
us; and the former, that by seeing the
gallant actions that have been performed, we may be excited to imitate
them.

There are no instances wherein military virtue shines with greater lustre,
than in those where seemingly our
weakness should cause our defeat.

To tire out a superior enemy, who
expected to have led us off in triumph;
to repel him, and throw upon him the
shame of a broken and ill concerted
project;—this is what characterizes the
great officer; this is the highest ability
a soldier can be master of.

Let it not be imagined that what I
say here, are only high sounding words
void of sense; the examples which I
shall cite will prove their reality; and
the means I shall propose, will prove
the

the facility of executing actions of this
nature.

However little the affiftance was that
I could find in the military authors
that I confulted, I have endeavoured to
omit nothing in this work, that may
ferve to make that part of war that
I treat of, underftood; and I have taken
much lefs pains to adorn it with wit,
than to furnifh it with principles which
may be relied upon, and rules that may
be eafily followed.

I will explain what the principles and
rules are, when I have given an idea of
geometry theoretical and practical, which
an officer muft neceffarily be mafter of,
in order to know how to make an entrenchment.

THE
SCIENCE
OF
MILITARY POSTS.

CHAP. I.

Geometry neceſſary for an Officer.

GEOMETRY is the nobleſt part of mathematicks; it is the ſcience of meaſuring all things that have perceptible dimenſions.

Officers of infantry being never employed to direct the conſtruction of

great

great fortifications, (that being only
the bufinefs of engineers) the geometry,
neceffary for them, may be reduced
into a very fmall compafs.

The *field fortifications*, which they
may probably direct, are fo fimple,
that the only requifite knowledge is,
to trace ftrait lines, horizontal, parallel, perpendicular, and curved ones,
and to underftand the connexions between them, in order to execute them.

But as the operation of drawing
thofe figures with compaffes and rule
on paper, is very different from that of
tracing them with the fathomed line
on the ground, I will fay a few words,
in order to explain them both.

Of the Point, and of the Line.

Plate I.
Fig. 1.

Geometricians call (a Point) the fmalleft thing that can be imagined; it is
confidered in mathematicks as indivi-
fible; that is to fay, having no dimenfions.

A line drawn by a rule from one
point to another, is called a right line,
as

as A, B; this line is the ſhorteſt that
can be drawn between the ſame points,
and is conſidered, like all lines in general, as a row of points placed ſide by
ſide, in a ſtrait or right line.

This line A, B, is called horizontal,
when it is ſo level, as neither to riſe
nor fall towards A, or B.

C, is called a perpendicular line, Fig. 2.
being drawn right up and down, ſo
that it neither leans towards one ſide or
the other, ſuch as a thread would
mark, having a ball ſuſpended from it.

To draw a line D, perpendicular to Fig. 3.
a ſtrait line E, F, the point from
whence the perpendicular is to be
drawn, may be out of this line, or in
the line itſelf.

If the point G is out of the line H,I, Fig. 4.
then from this point, as center, deſcribe
an arch, which ſhall cut the line in two
points, as L, M ; from theſe points L,
and M, and with the ſame diſtance, or
radius, deſcribe two arches interſecting
each other in one point N, then draw
a line from the given point G, through
the interſection of the two arches N,
which

which line will be perpendicular to the line H, I.

But if the point O, from whence the perpendicular is to be raised, is in the line itself S, T, then from this point, as center, describe a femicircle which may cut the line in two points P, Q, from which two points, as centers, you will describe the arch R, with the fame opening of the compaffes, and draw from the point O, the line R, O, thro' the points, of interfection of thefe arches, which will be perpendicular to the line S, T.

If, in the fecond cafe, the point from which you are to raife the perpendicu-
Fig. 6. lar, was at the end of the line V, X, you muſt then prolong this line beyond the point V, to defcribe from this point, as center, a femicircle, which may cut the line in two points, and do the reſt of the operation as above.

A line is *oblique*, when it leans one way or the other from a perpendicular.

Fig. 7. The *tangent line* (b) is that which touches another line in one point only, without

without cutting it; it is different from
a fecant line, which is that which cuts
another line.

A *curved line* is that which differs Fig. 8.
from a ſtrait line, in going from a point
A, to another point B.

A *mixt line* is that which has part of Fig. 9.
it ſtrait, and the other curved, as C, D.

A *ſpiral* is a curved line, turning Fig. 10.
round, and always widening its diſtance
from the center, as E.

Two lines F, G, H, I, are parallel, as Fig. 11.
they are equally diſtant from each other,
ſo that they would never touch each
other, though they were to be pro-
longed to infinity.

If from a given point F, you would
draw a parallel to the line H, I, deſcribe from this point F, and with a
ſpace taken at diſcretion, the indefinite
arch G, L, then from the point L, and
and with the ſame opening of the com-
paſſes, deſcribe another arch F, H, then
take on the firſt arch a part L, G, equal
to F, H, and draw a line to paſs thro'
the points F, G, which line will be parallel to the line H, I.

D Section,

Section, or intersection, is the point
through which two lines or arches pass
to cut each other, as M, M.

Fig. 12.

A *circle* is a figure contained within
a single line, from all parts of which
the center is equally distant, as N,O,P ;
every circle is supposed to be divided
into 360 parts, or degrees, so that the
semicircle is understood to be 180 degrees, and a quarter of a circle is 90.
This division by degrees serves to measure the angles.

Fig. 13.

The *circumference* is the crooked line
that describes the circle N, P.

A strait line Q, R, drawn from one
point of the circumference to the other,
passing through the center, is called
the *diameter* of the circle, because it divides it equally into two parts.

Fig. 14.

A *semidiameter* is a strait line drawn
from the central point of a circle to a
point of the circumference, as S ; this
line is also called a radius.

A strait line T, V, which divides the
circle into two unequal parts, is called
a *chord*, and the portion of the circle
X, cut by this line, is called an arc or
arch.
                                        Of

Of *Angles.*

An *angle* is a fpace bounded by two
lines which meet at a point, as A.   Fig. 15.

If a line B, is raifed perpendicularly Fig. 16.
on a ftrait line C, D, the angles that
they make are right angles; their meafure is a femicircle, that is to fay, they
have together 180 degrees, and each
of them 90.

Two right angles are drawn in the
fame manner as you raife a perpendicular in the middle of a ftrait line, as
aforefaid.

Likewife, to make one right angle Fig. 15.
4 A, the operation is the fame as is directed to raife a perpendicular at the
end of a ftrait line; its dimenfions are
ninety degrees, or the fourth part of
a circle.

An angle is *acute,* when the lines
that compofe it approach near to each
other, as E, and the angle contained Fig. 17.
between them is lefs than 90 degrees.

An *obtufe angle,* is that which is made
by two lines, which open from each
D 2   other

other to a greater diftance than a right
angle ; its meafure therefore is more
than 90 degrees, as F.

Fig. 18.

The top of the angle is that part
where the two lines meet in a point,
that compofe it, as G, H, I.

Fig 19.

To meafure the contents of an angle
of any kind, place one point of your
compaffes at the top H, and defcribe
with an indefinite diftance taken with
the other, the arch or portion of the
circle L, M, the dimenfions of this arch
which touches both lines of the angle
are the dimenfions of the angle itfelf,
let the opening of the compaffes, with
which this arch was defcribed, be what
it will; the reafon of that is, that whatever the opening of the compaffes was,
the angle G, H, I, will always be the
quarter of a circle. The fame rule
ferves for all angles.

Fig. 20.

To make an angle N, equal to another angle O : from the angle O, as a
center, with your compaffes opened to
any diftance, defcribe the arch P, Q;
then, with the fame opening of the compaffes, and from N, as a center, defcribe

scribe the indefinite arch R, S, then take the distance P, Q, the size of the given angle, and set it off from R to S, and draw through this point S, the line N, T; this line will form an angle equal to the given angle O.

Fig. 21.

It is so necessary in practice to know how to make one angle equal to another, that the execution of fortification, and all parts of the mathematicks, would be impossible without it; but what I have said serves for the construction of works sufficient to fortify posts in the field.

Of Triangles.

A *triangle* is a figure bounded by three sides, which form three angles.

An *equilateral triangle*, is that which has its three sides, and its three angles equal, as A, B, C. In order to describe Fig. 22. this, draw a first line A, B, then from the point A, and with the distance A, B, describe the arch D; then from the point B, and with the same opening of the compasses, describe another arch E; then draw through the point of inter-
seftion

fection (c) the lines A C, B C; the triangle formed by thefe lines will be
equilateral.

Fig. 23. A *triangle* is *right angled*, when it has
one right angle, as F.

Fig. 24. An *Ifofceles triangle* is that which hath
two fides, and two angles equal, as
G, H.

Fig. 25. A *Scalene triangle* is that whofe three
fides and three angles are unequal, as
I, L, M.

The three angles of any triangle are
equal to two right angles, that is to fay,
Fig. 26. the three arches N, O, P, defcribed
with the fame opening of the compaffes, make together 180 degrees.

The *contents* of the *furface* of a *right
angled triangle*, or of one that has a right
Fig. 27. angle, as R, is equal to half the product arifing from its height multiplied
by its bafe, or to the whole product of
half its bafe, by its height; becaufe it
is half of the fquare R, S, which would
have the fame bafe and the fame height.

Fig. 28. The line T, which is drawn from the
point of an angle, perpendicular to its
bafe, is called the height of the triangle;

angle; this line will form two triangles, which will of courfe be right angled, as V, X: you find the contents as above-faid *.

The doctrine of meafuring triangles is called *Trigonometry*; which is one of the nobleft parts of Geometry, and is treated at large by many excellent authors.

Of Surfaces.

Though the knowledge of *furfaces* and of *folids* is ufelefs in conftructing fmall field fortifications, neverthelefs I will explain them, as being an effential part of the principles of geometry.

Surface, or fuperficies, is a figure determined by many fides ‡.

A *plain furface* is that, which is even like a looking-glafs; it is *convex*, when it rifes in form of a globe; and *concave*, when it has any depth.

The

* An Equilateral, or Ifofceles triangle, fo divided, will make the two new triangles equal each to each; but in a Scalene fo divided they will be unequal, though always rectangular.

‡ Or in other words, a Surface is that which has length and breadth, for a circular Surface is bounded by one line.

The line being confidered only as a row of points, the furface is alfo confidered as a row of lines placed befide each other.

Fig. 29. A furface as A, whofe four fides are equal, and the four angles right, is called a *fquare*; the furface or area of this fquare, which is alfo called a rectangled fquare, is equal to the product of its bafe, multiplied by its height; that is to fay, if it is four toifes or fathoms long, and four toifes broad, its furface will be fixteen fquare toifes, or fixteen rectangled fquares, each of which will be a toife in bafe, and a toife in height.

Fig. 30. The *parallelogram* or *oblong fquare* B, is a figure whofe oppofite fides are equal, and the angles right; its furface or *area* is meafured like that of the rectangled fquare, that is to fay, if it be nine toifes long, and four toifes broad, its furface will be thirty-fix fquare toifes.

Fig. 31. A *rhombus*, or lozenge C, is a figure of four equal fides, but which has two oppofite

opposite angles acute, and the two
others obtuse.

The *trapezium* D is a figure, whose Fig. 32.
four sides and four angles are equal.

The figure E is called a *polygon*,
which has many angles, and above four Fig. 33.
sides : A *regular polygon* is that, which
has all its sides and all its angles equal;
by the term *polygon* is sometimes understood, the whole of a *fortified place*,
and sometimes only the ground traced
out in order to raise the works.

A *polygon* that has five sides, as E,
is called a *pentagon* ; *hexagon* that which
has six ; *heptagon* that which has seven ;
octagon that which has eight ; *enneagon*
that which has nine ; *decagon* that
which has ten ; *undecagon* that which
has eleven ; and *dodecagon* that which
has twelve.

When a place is fortified with six
bastions, it is said to form a *hexagon*.
It is regular, if its sides and its angles
are equal ; and irregular if they are
not.

E   All

All furfaces, except a *circle* *, that are not rectangled, are meafured by dividing them into rectangled triangles; of which I faid that the contents were, the product of half the bafe, multiplied by its height.

Of Solids.

A *folid* is a figure which has three dimenfions, viz. length, breadth, and thicknefs.

As lines are only rows of points, and furfaces only rows of lines, in the fame manner folids are only rows of furfaces, fuppofed to lie one on the top of the other, like the leaves of a book.

Of the different kinds of *folids,* the principal are, the *cube,* the *parallelopipede* or *parallelipede,* the *prifm,* the *cylinder,* the *pyramid,* the *cone,* and the *fphere.*

Fig 34.

The *cube* is a figure, whofe length, breadth, and thicknefs are equal, as F; a *die* is a cube. The contents of a cube

---

* The furface or area of a circle is found by multiplying the circumference by one quarter of the diameter. *See the following note on the fphere.*

[27]

cube are found, by multiplying the
length by the breadth, and the product
of those two by the thickness; that is
to say, if it be four fathoms long, and
four fathoms broad, whose product is
sixteen square fathoms, in multiplying
those sixteen by four of depth, you
will have sixty-four cubic fathoms, or
sixty-four solids, each of which will
have one fathom in every one of their Plate II.
three dimensions.

A *parallelipede* is a solid, bounded by
six parallelogram sides, whose opposites
are parallel and equal, as G. You find Fig. 1.
the contents of a parallelipede, by multiplying its dimensions one by the other,
like the cube.

The *prism* is a solid, which has an
equal thickness in its whole length, and
whose upper and lower bases are equal:
this name is particularly given to a triangular solid, as H, bounded at its Fig. 2
two ends by two triangles, equal and
parallel to each other, having three pa-
rallelograms for its sides, which cannot
be parallel to each other.

E 2   The

The *cylinder* is a round body, equally
thick in its whole length, and whose

Fig 3. bases are equal circles, as I; the contents of prisms and cylinders are equal
to the product, arising from the area of
their base, multiplied by their height.

Fig. 4. A *pyramid*, as L, is a solid, whose
base is square or triangular, and which
ends in a point; a pyramid is the third
part of a prism of the same base, and
of the same height.

Fig 5. A pyramidal figure, as M, is called
a *cone*, whose base is a circle, and whose
top ends in a point; its solid contents
are found, like those of the pyramids,
by multiplying the area of its base by
one third of its perpendicular.

Fig. 6. A solid that is round, as N, is called
a *sphere*, like a globe, or a ball. The
measure or *area* of the *surface* of a
sphere, is the product found by multiplying the circumference * by the diameter;

*, By knowing the diameter of a circle, the
circumference is found by the rule of proportion.
For as 7 is to 22, so is the diameter to the circumference; or if the circumference be given,
then reverse the proposition, and say, as 22 is to
7, so is the circumference to the diameter.

meter; and the folid contents of a
fphere or a globe, is the product found
by multiplying the area of the furface
by one third of its radius, *i. e.* one
fixth part of the diameter.

Of Practical Geometry.

The rules of practical *geometry* are
the fame as thofe of theoretical geometry, whofe object is to reduce the
principles to ufe, and to mark on the
ground, the different figures that may
be drawn on paper.

The neceffary inftruments to trace
geometrical figures on the ground are,
a *toife* or *fathom*, a *chain*, a *plain table*,
picquets, a *level*, and a *plomb*.*

In the place of the toife, which is a
wooden meafure of fix feet long, each
foot compofed of twelve inches, *&c.*
one may ufe a fword of three feet long,
which will be found fufficient. This is
a convenient length for a fword, and on
all

* But as thofe may be too many for every
officer to carry, I will fhew how their places may
be fupplied on all ordinary occafions.

all detachments an officer will have his
meafure with him.

In the place of the *chain*, which is
made of iron, and divided into five, and
five, or ten, and ten toifes, by a pen-
dant; one may ufe a line divided into
toifes by fo many knots. This fhould
be fixty feet long, and a loop handle
fhould be left at each end, the length of
which is to be included in the firft, and
in the laft toifes. This loop ferves to
pafs on the picquet that is to be drove
into the ground, from which the line is
to be ftretched. It is true that this line
will be fubject to fmall alterations by
being wet or dry; but field forts, not
intended to laft long, do not require a
fcrupulous exactnefs.

The *plain table* is one of the feveral
inftruments ufed to take angles in fur-
veying, but too cumbrous for an officer to carry, nor indeed is it neceffary
in what we intend to treat of here.

Picquets are fticks of about three feet
long, and of an inch or an inch and a
half diameter, fharpened at one end.
They ferve to ftretch the line, by driv-
ing

ing them into the ground, with a mallet
or great ftone, to draw the lines, and to
mark the tops of the angles of the intrenchments that are to be drawn : but
as wood is to be found almoſt every
where for this uſe, one need not carry
them ; but it is enough to know that
twenty of them are neceſſary for a ſquare
redoubt.

As the *level* may be omitted, I ſhall only
ſay that it ſerves as a guide to make the
earth even, and to draw horizontal lines.

The *plomb* ſerves to draw perpendicular lines in carpenters and maſons
work ; but it may be ſupplied by tying
a bullet or a ftone to the end of a ſtring,
which you may hold before you to the
work.

Such are the inſtruments uſed in
tracing on the ground ; thoſe which
are uſed after to dig the earth, are
ſhovels and *pick-axes* ; and to cut wood,
hatchets and *bill-hooks*, all which are ab-
ſolutely neceſſary, and it is impoſſible to
ſubſtitute any thing in their places.
Therefore a body of men ought never
to march to a poſt, without being provided

vided with one or two of each kind. I
now come to the method of tracing.

Fig. 7. To trace a ſtrait line from A to B,
I plant a picquet at the point A, on
which I fix the loop which is at the end
of the line, which I draw tight towards
B, with a ſecond picquet : having done
this, I mark with the point of a third
picquet, a track on the ground, gently
touching the line all along as I mark.

To raiſe a perpendicular line at the
Fig. 8. point C, of the line D, E, you muſt fix
a picquet at the point C as center, on
which you muſt hang one loop of the
line, and deſcribe, in turning, with the
point of another picquet (alſo faſtened
to the line, at the diſtance of a toiſe)
the ſemi-circle F, G, then from thoſe
points F and G, where picquets are to
be drove, you ſhould trace with the
diſtance of two or three toiſes, the
arches H, I, then fix the line again to
the picquet C, which draw tight towards L, in paſſing thro' the ſection of
the arches H, I, and draw the line
C, L, which will be perpendicular to
the line D, E.

If

If. this perpendicular M, should be Fig. 9. raised at one of the ends of the line N, O, you must prolong this line towards P, trace the semicircle Q, R, and finish as before said.

But if the point S, from which a per- Fig. 10. pendicular line is to be drawn, was out of the line T, V, you must fix a picquet at this point S, on which you must fasten your line by one of the loops, and trace the portion or segment of a circle intersecting the line T, V, in the points X, Y, then from these points X and Y, where you must fix picquets, you must trace with the line, with an equal distance, the arches Z, and make the line S, Z, pass thro' the point of intersection Z, and the given point S, which will be perpendicular to T, V. Such is the method to trace a perpendicular with precision; but as this operation is a little complex, and as one may not always have time to execute it, I believe it will be enough to see that the cord when stretched from *a* to *b*, in order to Fig. 11. trace the perpendicular required, forms,

F   23

as near as the eye can judge, two right
angles with the line *c*, *d*.

To trace two lines parallel, you muſt
after having drawn the firſt line *c*, *b*,
meaſure the diſtance to which you would
trace the parallel, which I ſuppoſe to *g*,
deſcribe from this point *g*, as center,
(where you muſt fix a picquet) the indefinite arch *h*, *i*, and from the point *h*,
deſcribe the arch *e*, *g*; then take on the
firſt arch the part *h*, *i*, equal to *e*, *g*;
and laſtly, trace the line *g*, *i*, this ſecond line will be parallel to the line *e*, *h*.

But as theſe works, which officers may
have occaſion to make, are of but little
extent, and that great nicety is not required, one may ſhorten this operation,
by meaſuring towards the two ends of
the line *e*, *h*, with a ſword of the length
before-mentioned, the two equal diſtances *e*, *g*, and *h*, *i*, as perpendicular
as it is poſſible to judge by the eye, and
trace afterwards thro' the marked points
the line *g*, *i*, which will be parallel to
the line *e*, *h*.

When I ſhew how to trace ſquare redoubts, I will explain the ways to trace

Fig. 12.

a

a triangle, a perfect square, and a circular figure; but I will not speak of regular or irregular polygons, becaufe
thofe figures are only ufed in the conftruction of great fortifications.

As the operations which require the
tracing ftrait lines, perpendiculars, para-
lels, and angles, are thofe which are
ofteneft neceffary in field fortifications,
young officers ought to practife them
often.

There is often a great deal of leifure
time in a fettled camp, as well as in
garrifon, in time of peace, which might
be employed in thofe amufements,
where willing foldiers may be made to
affift, by giving them a fmall recompence; I fay by foldiers rather than
peafants, becaufe this advantage will
accrue; that is, they will work cheaper,
and it will accuftom them to work with
more eafe when the war may require
their fervice.

F 2   CHAP.

CHAP. II.

*Of the different Works with which Posts
may be fortified.*

IT may be expected that I should
speak of the *detachments* to *posts* before the manner of intrenching them;
the detachments being always ordered
to work thereon; but the drawing geometrical figures, having led me towards *intrenchments*, I think them proper to follow.

The security of an army depends on
the *defense* of its *posts*, and on the vigilance of the detached guards.

Let the abilities of the General be
ever so great, it is impossible that he
can have an eye to all the little details
that contribute to their defence; it is
sufficient if he knows that the guards
are well posted, and that the line they
form be well supported. It is afterwards the duty of the several officers
that command them, to make the best
dif-

difpofitions for a vigorous defence, that
they may anfwer the General's views.

An officer is detached to a poft, either to relieve a party, or to take firft
poffeffion of it himfelf: In the firft
cafe, it often happens that the guard to
be relieved is intrenched ; as foon as he
arrives at the poft, and has taken the
charge thereof from the officer that
commanded, he is to prepare for his
defence, as I fhall explain on this article. In the fecond cafe, if an officer
that is detached would intrench himfelf, he muft obferve, firft, to chufe the
place, to throw up his intrenchment, fo
that he may from thence difcover all
approaches ; for if the enemy could
come, without being perceived, to within a fmall diftance of his poft, the af-
failant might cover himfelf and his
party, where he might remain in fafety,
and keep the befieged always under
arms, taking time to advance upon
them whenever he pleafed

If on the ground, where he would
throw up his works, there may be any
hollow ways, a thicket of wood, or any
other

other means of the enemy's being covered, they muſt be rendered uſeleſs to
the enemy, or guarded by detachments
of ſix or ſeven men. Secondly, He
ſhould take care that there ſhould be no
high place near that may command
him ; or he muſt hinder the enemy to
profit by it : becauſe if the enemy
ſhould take him in flank, or fire into
the work, it would be impoſſible for the
ſoldiers to defend it. The manner how
to avoid this inconvenience ſhall be explained hereafter, when from the nature of the ground there is no taking
poſts to avoid thoſe heights. Thirdly,
He muſt make his work in proportion
to the number of his men who are to
defend it.

Good ſenſe and various examples
ſhew that too large intrenchments, ſuch
as are frequently made, cannot be well
defended but by a conſiderable number of men. An exceſs of this kind
ſeems to me a great fault. I think it
better to fall into the oppoſite, in making them ſmaller. Fourthly, He muſt
be attentive to give equal ſtrength to
all

all parts of the work, that it may refift the enemy alike every where. Fifthly, and laftly, He muft try ftrictly to fulfill the intended fchemes of the General, by maintaining his poft.

If he is detached to an open countay, or to a height, and that he may be attacked on every fide, as is often the cafe with fmall guards, he muft build a redoubt, or fmall fquare fort, with a parapet, a banquet, and a ditch.

When the ground is chofen, he fhould mark a ftrait line a, A, E, and raife the perpendicular a, B, C, as defcribed in the article of Practical Geometry; fet off from a towards c, and from a towards E, the dimenfions propofed for each fide of the parapet within the fort, which, if his party confifts of thirty men, fhould be two fathoms or two and a half, four fathoms for fifty, and eight fathoms for one hundred men : this will afford a fpace oppofite the parapet of almoft two feet for each man This fpace no doubt will appear too large for the defence of an intrenchment, where the men are to be drawn

Fig. 13.

drawn up at leaft two deep; but it is impoffible, neither is it ever the practice, to make their proportions lefs, except only when the detachments are very numerous; the fize of the parapet may be fo proportioned, as to admit the men who are to line it to be drawn up two and even three deep.

After having drawn thofe two firft lines A and B, as before faid, the loop of the cord is to be fixed or hung on the picquet C, of the perpendicular line B, and with the fame length a, c, trace the arch D; then hang the cord on the picquet E, at the end of the line A, and with the aforefaid diftance, or length, defcribe the arch F; the point of interfection in thofe arches determines the length of the lines E, H, and C, G; thofe four lines fo drawn form a fquare, which will mark the interior fides of the parapet.

Then four other lines I, L, M, N, are to be drawn at the diftance of two or three feet within fide this fquare, and parallel to the firft, to mark the breadth of the banquet, which is to be

more

more or lefs, according to the intention
of drawing up the men, who are to
ftand thereon, in more or fewer files.

In the next place, a third parallel
fquare, O, P, Q, R, is to be drawn
without-fide the firft, to determine the
exterior fide of the parapet, and its
thicknefs, which is commonly eight or
nine feet; but if the intention is to refift cannon, it muft be eighteen feet,
or three fathoms.

Laftly, a fourth fquare, S, T, V, X,
muft be drawn to determine the breadth
of the ditch, which muft be the fame
with the parapet, or may be two feet
more than the thicknefs of the parapet,
and a picquet muft be fixed at every
one of the angles, as well as at thofe
of the lines already marked, to avoid
lofing the points of the fquare.

While the drawing and marking are
executing, with the help of two or three
men, five or fix more fhould be employed in cutting down the neareft trees
to the poft, as well to difcover the approaches of the enemy, as to ferve in
making the intrenchments : the fmall

G   branches

branches ferve for fafcines, which are a kind of faggots about a fathom long, and two feet thick, uniform from end to end ; they are bound at each end, and in the middle, and ferve in the intrenchments to keep up the earth, which, without them, would crumble down ; the middling branches of thefe trees ferve to make ftakes, or picquets, to drive between, or through the fafcines, to faften them to the ground, or one on the top of the other, in order to raife the parapet. Laftly, the trunks and greateft branches are ufed to ftrengthen the poft, as fhall be fhewn hereafter.

The drawing being finifhed in the above manner, the firft range of fafcines is to be laid on the fmalleft fquare, I, L, M, N, as a foundation to fupport the earth of the *banquette* ; then the fecond range is to be laid on the fquare A, B, G, H, to keep up the infide of the parapet ; then the third range on the fquare O, P, Q, R, to keep up the outfide of the fame parapet.

You

You muſt remember, when you are pinning down, or picquetting your firſt faſcines, to leave on the ſide leaſt expoſed to the enemy, a ſpace of three feet, for entrance into the redoubt ; but if it be in the power of the enemy to get round, ſo as to fire right thro' or *enfilade* this paſſage, then let it be made winding like the figure Y, in which form it Fig. 14. cannot be enfiladed.

After having picquetted the three ranges with faſcines, as I ſaid before, you muſt dig the ditch A, B, diſtant a Fig 15. foot from the outſide of the parapet ; this ſpace, or breadth, which is called *berme*, ſerves to keep up the mould, and to receive the rubbiſh thrown down by the enemy's cannon from the parapet : this *berme* is broader, or narrower, according to the ſtiffneſs of the earth ; throw the mould into the intervals C, D, E, marked for the parapet and the *banquette*, and make the men tread down the earth, ſo as to make it hard ; alſo take care, in hollowing the ditch, to leave a *talus*, or ſlope, more or leſs, according as the nature of the

G 2   ground

ground requires, to both sides F, G, to prevent the earth from falling down.

The slope F, which is next the redoubt, is called the *scarpe*; and that on the outside next the field, as G, is called the *counterscarpe* : You must take care in picquetting the fascines, as you raise the parapet, to bring the fascines of each face a little closer together, as you see at H, so as to leave the same slope of each side the parapet. The space E, D, marks the *banquette*, the space D, C, the thickness of the parapet below, the space I, L, the thickness of the same parapet at the top ; the space M, N, the breadth of the bottom of the ditch ; and the space A, B, the breadth at the top of the ditch.

You are to raise the *banquette* of this work two feet, if the ground is even ; but if there are some places too low, you must make two *banquettes* one over the other with steps. You must make the parapet four feet higher than the *banquette* ; but if this *banquette* was risen on account of some neighbouring height,

height, from whence you might be enfiladed, you muſt alſo raiſe the parapet,
till you ſee the enemy cannot hurt
you.

Leave to the upper part of the parapet a *talus*, or ſlope, I, L, that your
men may ſee all without, and to fire
well towards the country O.

Tho' the form of a ſquare redoubt,
as I have deſcribed, may be what is almoſt always uſed in the field, it has
neverthelefs defects, which ought to be
an objection to its uſe, at leaſt on poſts
that are to be defended on all ſides
alike.

Experience ſhews us, that the defence, by oblique firing of muſketry,
is not to be depended on ; for a ſoldier
hardly ever fires in any other direction
than right forward, as A, and even
without aiming. By this way of firing Plate II.
it happens, that great ſpaces, oppoſite Fig. 16.
the angles of the redoubt, as B, are left
without defence, where the enemy may
remain in ſafety.

The Chevalier de Clairac, an experienced and good engineer, gives, in his
treatiſe

treatife of flight or temporary fortifications, an excellent way to remedy this inconvenience ; it is to make the inner fide of your parapet indented, or as it were a row of little redans, fit for one or two men of a fide : this method is Fig. 18. the more excellent, as the crofs-firing takes the enemy on each flank, and there can be no approach that is not defended. But this kind of redoubt has too much work in it, and takes up too much time in conftructing, to be made by regimental officers ; I would join with the fame author, to make cir-Fig. 17. cular redoubts C, becaufe every point of the circumference being equally difpofed, the foldiers may ftand any where, and the exterior fpaces D, which are defended, changing every minute, the enemy has no place of fafety.

The circular redoubt therefore is the moft perfect that can be made ; but when you have a road, or the bank of a river to defend, the fquare redoubt, the oblong, or triangular, are preferable ; becaufe you ought to oppofe the faces of your intrenchment as low as
poffible,

[47]

poſſible, and parallel to the parts that you would attack, always obſerving to round off your angles.

In order to draw a circular redoubt, after having choſen your center, you muſt fix a picquet at the center point; and from this point, with a determined length of line, according to the number of your people *, deſcribe the circle E, E, to mark the inſide of the para-pet; then draw another within that, allowing the breadth that I ſaid before for the *banquette*; then draw a third F, F, to mark that of the parapet; laſtly, draw a fourth G, G, to give the outſide breath of the ditch : this done, picquet your faſcines, in giving them the curve line of the circles, and finiſh as in the ſquare redoubt.

Fig. 17.

If

* If you have thirty men, give a fathom and a half to your cord, for the length of the radius, which will make three fathom diameter, and about nine fathom circumference, and a little leſs than two feet for each ſoldier. If you have fifty, you give two fathom radius, which makes four fathoms diameter, and about twelve fathoms round. Laſtly, if you have one hundred, you double theſe proportions, unleſs you mean to draw up your men two or three deep.

If an officer commanded a detachment that was pofted on the pafs before
a bridge, in a defile, or before a ford,
he might make a parapet in a ftrait
or a curve line with its *banquette*, and
a ditch that fhould ftop all the entrance:
this would be better than a redan,
which is a work that has but two faces.
I will not defcribe this work here, becaufe one officer never has men enough
to defend the parapet, which is commonly very extenfive. For the fame
reafon I will not fpeak of ftar forts, or
the larger forts, where only great detachments will do.

I will defcribe the ways to fortify
houfes or villages, when I come to fhew
the manner of adding fortifications to
fuch pofts as are already ftrong by nature.

C H A P.

CHAP. III.

*Of the different ways of encreasing the
strength of Posts.*

IT is not only with the works that I
spoke of, in the former chapter, that
an officer may fortify his post; there are
also an infinite number of ways how to
stop, to tire out, and even to repulse
the enemy, of which he should not be
ignorant.

The strength of a redoubt A, or of Plate III.
any other work, may be encreased by
filling the ditch B with water, which is
done by turning the stream of a spring
or rivulet into it, or by cutting a drain
from a river or pond. But if the ground
of your post is uneven, which sometimes happens, and that you cannot run
the water equally into all parts of the
ditch; you must, in hollowing it, make
batardeaux or dams, as C, or little tra-
verses of earth, which make proper
banks to keep up the water in the
higher parts of the ditch D, from whence

H you

you may let it run down into the lower, E. You muſt allow to theſe banks at the top D, only half a foot thickneſs, which will be pretty ſharp; but you muſt leave it much thicker at the bottom E, giving a great ſlope to each ſide.

Dams, like F, are alſo made with boards or planks; but they muſt be ſtrong, and kept up by great ſtakes, ſo that the weight of water above ſhould not puſh them down: This kind of dike is preferable to thoſe of earth.

You may alſo ſtrengthen your works, by ſtopping up or embarraſſing the environs and the avenues that lead towards you. In a mountainous country you may interſect the roads with large ditches, or break them ſloping; you may ſtop up the defiles with waggons one a-top of the other, and poſt a few muſketteers behind them; laſtly, you may * lay heaps of briars and thorns half buried in the ground, wherever

you

* I think if roots of trees, or large branches or blocks of wood, are placed, part of them buried, they will do better.

you find it neceſſary. But if, as it often
happens, the General orders the officer
to retreat towards the army, or to fall
back to another poſt, if he ſhould be
attacked; he muſt take care not to de-
ſtroy the road ſo as to hinder his own
retreat, but muſt leave himſelf a paſſage,
made like a draw-bridge, or ſome other
way, which muſt be guarded by ſeven
or eight men.

If he is detached to a plain or flat
country, he muſt dig deep ditches in
the avenues to it, and in the approaches
to the poſt; or pits, which he may
cover afterwards with ſlight boughs,
and a little earth over them, and take
care to ſpread the earth dug out of
theſe pits on each ſide, that the enemy
may not know exactly where they are.
Alſo he may ſcatter in the avenues
caltrops, which are, as it were, iron
ſtars with four ſpikes, ſo diſpoſed, that
whatever way they are thrown, one ſpike
will always ſtand pointing upwards.
Laſtly, he may fix picquets quite round
his poſt, very near each other, and a
little inclining to the field outwards,

about

about two feet out of the ground, and
sharpen their points afterwards.

But the greatest obstacle that he can
oppose, is what M. le Chevalier Follard
depends most on, in his Commentaries
on Polybius; which is, to shut up the
roads, and embarrass the defiles, and sur-
round your post with a breast-work of
trees, with their trunks buried about
three or four feet in a ditch made on
purpose. The trees fit for this use ought
to have large branches, and you may
sharpen their points, and take off all
the leaves; place the trees as near each
other as you can, so that the branches
may twist into one another, and see that
they point a little towards the enemy.
You may make, if you chuse, three or
four ranges of breast-works of trees
round your redoubt; but they must be
at two fathoms from each other, that
the enemy may not burn them at the
same time, to approach your redoubt.

" Good redoubts," says M. de Saxe,
in his Reveries *, " are the most ad-
" vanta-

* The pocket edition of Reveries of M. de
Saxe, pag. 326.

" vantageous, as they are the fooneft
" made, and they ferve on very many
" occafions, where one only, in a clofe
" country, will fometimes ftop a whole
" army from annoying you on a critical
" march, and enable you to occupy a
" deal of ground with a few troops."

Sometimes you may cover your party
by a fimple *abatis* (or breaft-work of
trees) when you do not intend, or have
not time to throw up one of earth; and
you muft take care to place the trunks
one on the other, as much as you can,
to make a kind of clofe parapet; otherwife the enemy, by forcing their way
to the breaft-work, and having open-
ings to fee even from the feet to the
heads of thofe within the work, may
kill them one after another.

If it is a ford, or a river, that you
have to defend, you may make a parapet, minding to make it as near the
water as you can, fo that the enemy may
not have ground to draw up, when they
have paffed over. You may add to the
difficulty of the paffage, in digging a
deep ditch before the ford, and letting
the

the river-water into it; you may alſo
ſlope down the bank ſides, throw trees
acroſs each other, and lay caltrops.
But all theſe devices, I have ſhewn, only
ſerve to encreaſe the force of the out-
ſkirts of your poſt, which are next the
enemy; there are other means that you
may uſe beſides what I have ſpoken of,
in places of ſome natural ſtrength, as
chateaux, chapels, farm-houſes, or barns.
An officer who is ſent to a poſt
of this kind, that ſtands by itſelf,
muſt take care, before he begins to
work, to oblige the inhabitants to
quit it, and lodge them, by the help
of the magiſtrate of the next village,
in ſome other place. The next thing
to be done is, to make a curving
parapet, if he has people enough to
defend it; if not, and if he has but a few
men, he muſt make a breaſt-work round
the houſe with trees, and chiefly before
the angles, to prevent the enemy from
undermining it there. He muſt alſo
take off the tiles or ſlates from the roof
of the houſe, ſo that the enemy, by
putting ladders againſt the walls, may
not

not knock thofe within on the head.
If the houfe is thatched, or covered
with any other matter that is combuftible, it muft be taken off, and
burned, left the enemy fhould employ
it to demolifh the houfe itfelf; and he
muft deftroy all other things of the like
nature, for the fame reafon. This is
the advice of M. Follard.

Though you have furrounded the
houfe with a parapet or *abatis* (i. e.
felled trees) you muft break fmall loop
holes through the wall of the ground
floor, and let them reach to within a
foot of the ground, fo as to difcover
the enemies legs, and to hinder them
from taking your outworks, by your
placing five or fix mufquets therein.

Thefe holes, which may be about
four inches wide, ought to be broke
through at about three feet diftance
from each other; and there fhould be
a little trench dug, at a foot and a half
from the wall, and within-fide the houfe,
wherein the men fhould be placed to
defend it.

Alfo

Alſo make other holes, ſeven or eight feet from the ground, oppoſite the intervals of thoſe below, and of the ſame ſize; and let your ſoldiers that are to defend theſe ſtand on benches made of boards, or planks, or tables, or ladders; and take care to make a great number of holes oppoſite the avenues leading to the doors, or the angles of the houſe; becauſe here you may expect the ſtrongeſt efforts of the enemy. If there is a court to the houſe, you muſt make holes through the walls that look into it, ſo as to be able to warm the enemy, if they ſhould get into it.

If there are many doors, you muſt barricade them, and ſtop the paſſages towards them, by laying four or five trees one on the top of the other, leaving to the one that is to be the entrance of your poſt, only room for one man at at a time to come in. If there are low windows, that are not defended with iron bars, you muſt ſtop them up with dung, planks, ſtones, earth, and even trees.

If

If the houfe has large offices on the
ground floor, or the like, you muſt
bury trunks of trees in the middle of
them, leaving the branches ſticking up,
to hinder the enemy from forming, if
they ſhould get into the houfe. Laſtly,
you muſt be attentive to plant one or
two of thefe trees three or four feet
within the entrance port or door, to hinder the enemy from forcing in front-
wife, and to lay them under the difficulty
of ſqueezing in fide-ways

If there are great ſtairs in the houfe
to go up to the firſt ſtory, you muſt
break them down, or ſtop them up
with ſtones, or caſks filled with earth.
If the ſtairs are in a wing projecting
from the houfe, as it often happens,
you muſt break holes through the walls,
to fire on the enemy that ſhall have got
in ; and you muſt ufe ladders to get
up into the firſt ſtory. Alfo make feveral holes, through the floor, of four
inches diameter, to fire from above on
the enemy below. You muſt not make
thefe holes in the boards over the place
where you have fixed the trees, but

I   make

make many over the door, and at all
the weakeſt parts, where the enemy are
likely to force in.

You may alſo break holes through
the walls of the firſt ſtory, about three
feet from the floor, and leave the holes
at leaſt ten inches wide, and let them
be broke three feet aſunder, and over
the intervals of thoſe of the lower
ſtory.

As to the windows of this ſtory, if
you have not people enough to defend
them, you muſt ſtop them up, to hinder the enemy from placing ladders, to
fire in on you. You may, ſays the
Chevalier Follard, make a great opening, before each window, in the floor,
ſomewhat broader than the window,
which will be a kind of ditch, into
which thoſe that ſhould attempt the
window muſt tumble.

You may make the ſame prepara-
tion in the ſecond and third ſtories
as in the firſt, that if the enemy would
undermine you below, or break in
above, they may find an equal reſiſtance
every where ; but it will be uſeleſs to
make

make holes in the top story, as you
may, (having taken off the roof) take
down the wall to breast high, and fire
over it; referve the stones or bricks,
and place them in heaps, to throw
down on the enemy, and the rafters,
to beat down the ladders that may be
applied to the wall.

A post intrenched in this manner,
may hold out a long time, and even
tire out the besiegers, if it be defended
by resolute soldiers; and such are always to be found in an army.

M. D'Enfernay, who was very ex-
pert in fortifying posts, such as I mentioned, was detached, in the campaign
of 1748, to Bevera, a village on the
west side of Genoa, two leagues from
Vintimillia on the Roya, with a free
company, which he commanded: he
took post in the church of the place,
which stood by itself, and encompassed
it with a parapet and wet ditch; but
part of his intrenchment was commanded by some houses of the village,
so that the enemy might fire down on
his people, and take them across the

I 2   para-

parapet. He remedied this defect, in covering the part that was commanded with a kind of blind made of rafters, one end leaning again the church, and the other end fupported by pofts a foot higher than the top of the parapet, which gave his men liberty to fire under it ; and it being covered with fafcines and earth, the fhot of the enemy could not hurt them within, who could however pelt thofe without.

I was detached from the army at this time with a body, to act under the orders of this experienced partifan ; and I could not help admiring the defences that he contrived to his poft, where the enemy did not dare to vifit him, tho' they were his neareft neighbours.

I thought it my duty to mention this example, as well to do juftice to that officer, as to fhew the method of covering yourfelf in a poft that is commanded by a height. As for thofe that have no natural ftrength, fuch as redoubts, and other intrenchments of earth, this defect is remedied by raifing the fide of the parapet that is commanded

manded, as I faid before, or in making a fcreen with rafters or poles fixed perpendicularly againft the infide edge of the parapet, to which you may nail afterwards fome planks, or fafcines, obferving to leave a fpace of half a foot between the top of the parapet and the bottom of the blind, for the mufquetry to fire through.

But if an officer has not time for all thofe works that I have fpoke of; which happens when a General has a mind to forage, and throws fome foot into the houfes, or farms, to form a line; an officer ought then to lay directly two trees acrofs before the door, and cut holes through the floors, ftop up the windows, and prepare for a vigorous defence, which will give time to the foragers to retire, or to detachments to arrive to fuccour them.

What I have faid only relates to pofts that ftand alone; but if an officer fhould have a village to defend, he may cut out a more difficult work for the enemy : when I fpeak of intrenching a village, I mean only to fpeak of thofe where
the

the houfes' ftand very clofe together, or of fuch as are fometimes enclofed with walls. A commandant fent to a poft of this kind, ought, before he begins to work, to go feveral times round it, to examine the approaches, and the houfes that are near it. Then he muft make holes in the walls of fome, as before faid, and muft block up the fronts of all fuch as have paffages leading to the fields, with *abbatis* or felled trees ; and if he has time, he may make a good *abbatis* quite round them, and intrench the entrance of the ftreets.

An officer that has a mind to fortify a poft of fome extent, in this manner, fhould fketch out a plan of the village, and of the intrenchments he intends to make, becaufe that will give him hints how it may be defended, which, on the view only of the country, might efcape him.

A ftreet is to be defended like a bridge ; that is to fay, with a *redan*, or rather with a femicircular fimple parapet, with a ditch. Break loop-holes in every ftory of the houfes, as before defcribed, which are near the entrance; make deep trenches

trenches or cuts acrofs the ftreets, and lay caltrops in the bottom; alfo obftruct the ftreets with trees, carriages, or cafks; likewife open many paffages from the backs of the houfes, to go from one ftreet to another; and take care, efpecially if you have few troops, to fill up the middle of every wide place in the village with many trees, fo as to hinder the enemy from having room to draw up there, if they fhould get in.

As cannon or fire are the things moft to be dreaded in the defence of a village, an officer ought to break up the roads, to hinder their arrival; which is very eafy to do in a mountainous country; but when the village is in a plain, he muft cut deep trenches here and there athwart the avenues, and lay feveral trees acrofs, fo as to cover the whole breadth. If he has time, he may prop up the great beams of the houfes in the out-fkirts, with trunks of trees, or large pieces of wood put up like ftones: this is proper, left the floor fhould fall down by a cannonade, and crufh the men below.

In

In regard to fire, he had beſt burn
all combuſtible matter, to prevent the
enemy's turning it to his deſtruction;
but if there ſhould be a great quantity
of wood, ſtraw, or hay, he ſhould firſt
aſk the conſent of the General thereto;
or, if he judges it neceſſary, to have it
carried off for the uſe of the army.

Another eſſential thing for an officer,
who is detached to a village, to be attentive to, is to ſecure himſelf a laſt retreat, in caſe he ſhould be forced in
the ſtreets, and at his firſt intrenchments. For this purpoſe he ſhould
chuſe the *chateau*, (or *manſion-houſe*) the
church, or any good houſe that ſtands by
itſelf, which ſhould be intrenched with
care, after having diſlodged the inhabitants.

When ſoldiers who defend a poſt
know that they have a ſtrong-hold to
retreat to, they never think of giving
up, or ſurrendering themſelves priſoners, while they ſee they are in a condition to obtain honourable terms of
capitulation.

But

But if the houses of the village to be
defended, are scattered, or if there are
gardens or courts in the center, you
must then confine yourself to intrench a
single house, or the church, or church-
yard, or the *chateau*, where you may
cover yourselves by an intrenchment of
earth, and with all the other little ar-
tifices, that I mentioned for places that
stand alone, observing always to take
care of your flanks.

As the different works to fortify a
post, and especially a village, are too trou-
blesome, and even too tedious, if they
were only to be made by the soldiers
of a detachment, who ought never to
be over-fatigued ; an officer ought to
command (with the assistance of the ma-
gistrate of the place) a certain number
of peasants to work alone, or together,
with a third part of the detachment,
who should all be provided with shovels
and pick-axes ; while the other two
thirds should remain under arms for fear
of a surprize. These soldiers and peasants who work, should be relieved
every three hours, and care should be

K   taken

taken to see that they have no concealed arms, and that they continue their labour, without interruption, till the end of the work.

During the war in Italy 1747, I employed the inhabitants of the village of Berra, in the county of Nice, in this manner, where I was detached with thirty men. M. de Mirepoix, General of the French army, in the absence of Marifchal Belifle, receiving advice, that some Piedmontese peasants had penetrated into this county, with a design to carry off some of our quarters; this General gave orders to all officers, who commanded the quarters, to be alert, and to send detachments to drive them back. In confequence whereof, M. de Charle, who commanded at the village of Contes, and in the diftrict of Berra, wrote to me to put this poft, one fide of the court of which was quite open, into the beft condition of defence that I could. The fame day a fpy of M. de Mirepoix came to my poft, and told me, that two companies of one hundred men each, who were marching towards Berra,

Berra, would arrive there before midnight. On fo exact advice, I ordered, by means of the magiftrate, thirty peafants, who broke loop-holes in the walls of the *chateau*, wherever I directed them, and raifed a good parapet of dry ftones at the opening of the court yard, where the enemy might have entered by (ladders, or) *efcalade*. Then I made them throw feveral trees, with their branches on, before this parapet; and fending for the magiftrate and his eldeft fon, a lad of about fourteen years old, I told the father, that being informed by the orders I had received, that the enemy, who were marching towards my poft, held intelligence with the inhabitants of the village, in order to carry me off, I would keep his fon with me, to fix him on the top of the parapet; and that the firft fhot the enemy fired, fhould be at him. Having taken thefe precautions, I fent the father back, and held myfelf ready to receive the Barbetts; who hearing that we had parties abroad to attack them, fell back to Tende.

After

After defcribing the manner of
ftrengthening a poft, it follows in order,
to fpeak of the detachments that are to
defend them.

C H A P. IV.

Of the neceffary preparations to go on detachment.

DEtachments are particular bodies
of foldiers, fent from a greater
body, to fecure a poft, or to go on fome
expedition.

When an officer is commanded to
go on detachment, he fhould put on
his gueters, fafh,* and gorget, and take
his fufie or fpontoon, and provide himfelf with a line, divided into fathoms,
to meafure out his intrenchments, if he
finds them neceffary to be made.

He muft be early on the parade or
place of rendezvous. When he is come,
and is told what party he is to command,

---

* Englifh officers alfo take their fafh.

mand, he fhould afk the Brigade-Major,
whether he is defigned to relieve another detachment? Or if he is to be the
firft to take poffeffion of the poft? If
he is to relieve a body, he is only to require a guide to conduct him. This
guide is a foldier fent by the officer to
be relieved; who goes as an orderly man
to the Adjutant-General, to be ready
to carry any orders that may be neceffary; and who having been at the poft
before, comes to conduct the new detachment. In cafe the poft is to be
taken for the firft time, the officer is to
afk the Brigade-Major for inftructions
relative to its defence.

Having taken thefe inftructions, he
muft examine the men of his party, and
be careful that every foldier is properly
equipped; to fee that his piece is loaded,
and frefh primed, that his flint * is good,
that

* The tranflator has obferved the Englifh flints
fometimes to be very good, and at other times
very bad; and what he has feen of the French
flints were uniformly good. He thinks a foldier
ought always to be provided with a fpare flint, or
more. This inattention in regard to flints in a
British

fire.

that their ſtoppers are taken out, that their pouches are filled with ammunition, and that the reſt of their accoutrements are in good order; and ſee that each ſoldier has his canteen and his bread for twenty-four hours, in which time they are commonly relieved: and never let a man be abſent, either to eat or drink till relieved; and take care that his men carry proper tools to intrench with; ſuch as ſhovels, bill-hooks, hatchets, and pick-axes, one or two of each ſort; and if he wants any thing, he muſt apply to the Major of the regiment.

Some young officers may ſay, perhaps, that theſe are precautions that nobody thinks of taking; but are thoſe then that I have ſhewn, ſo little eſſential as to be neglected? And is it not reaſonable to think, if a ſoldier is unprovided of theſe things, that he will conſequentially

Britiſh army, was very remarkable at the ſiege of Louiſbourg; where it was a very uſual phraſe among the common ſoldiers in a ſkirmiſh, to ſay, "Ha! Monſieur, if our *flints* were as good as "yours, we ſhould ſoon make you ſlacken your "fire."

fequentially be incapable of making the defence that he ought ? It is to no pur- pofe to fay, that foldiers are of courfe fupplied with every thing neceffary of this kind in the field; for, on the con- trary, I have often feen them wanting entirely every thing, fo as to become ufelefs members, and rather a burthen in a poft.

"This, fays M. de Vauban, is what "makes us fo unfuccefsful in defence; "owing to the neglect of many officers, "in the provifion of tools neceffary on "thefe expeditions : and the fource of "this neglect, which is too common, "befides their ignorance and impru- "dence, is, that they treat it as too "trifling an article to merit their atten- "tion; tho', in effect, it is one of the "moft important things to be ob- "ferved."

As to war, when you form a plan of a good defence, it is better to take a thoufand ufelefs precautions, than to neglect one good one; becaufe the leaft neglect may difconcert the beft mea- fures. But you muft never be dif- heartened

heartened by imagining the enemy more
vigilant than they really are, and by
ſtarting difficulties that moſt probably
never will happen.* If in war you ſtop
at every ſuppoſition that the imagination ſuggeſts, you will neither undertake nor execute any thing. One general rule in military projects is, never
to forget any thing that may make us
ſure of ſucceſs, and whereof the execution depends on ourſelves only : as for
what depends on the enemy, ſome part
muſt be left to chance.

When the officer has examined his
party, he may aſk his guide the nature of
the roads ; if they be narrow, wide, open,
or woody ; if the enemy's poſts are near ;
if they ſend out patroles ; if he ſees
their parties in the country in the day-
time : laſtly, if he is to paſs before any
cottages, farm-houſes, or *chateaus* ;
on theſe informations, the officer takes
the neceſſary precautions for his march
from camp.

C H A P.

* See the accounts of the expedition to Roche-
fort in 1757, and that to Louiſbourg in 1758.

[73]

CHAP. V.

Of the march of detachments to pofts.

THE march of a body of men to a
poft, is one of the moft important
duties on which an officer can be employed in time of war.

I will not fay here, that night is the
propereft time to perform this march,
becaufe the time of their departure is
fixed by the General himfelf: I will
only obferve, that there are certain neceffary precautions, which ought never
to be neglected.

If the poft, to which the detachment
marches, is diftant from the camp, the
officers fhould not get on horfeback till
they are out of fight of it, and fhould
difmount when they come in view of
the poft, and have the horfes led back
by their fervants; but if the place to
which they are detached, is but a league,
or thereabouts, from the army, and
near the enemy, I think they had beft
go on foot, in order to be lefs em-

L   barraffed,

barraffed, in cafe of any fkirmifhes on
their march.

But whether they go on foot or on
horfeback, they, as well as their ferjeants, fhould take great care not to
hurry the men too faft, left fome fhould
not be able to keep up; to march clofe,
and in as many files as the roads will allow; never to ftop, and to be very filent,
fo as to hear all orders that fhall be
given. You may fee in a treatife attributed to Marfhal Saxe, entitled,
Traite des Legions, ou Memoires fur
l'Infanterie, printed in 1753, of what
confequence it is to a whole army, and
to an officer particularly, to march in
good order. The paffage is as follows:

" All the armies that the King fent
" into Bohemia, Weftphalia, and Ba-
" varia, marched off well equipped, in
" fine order, and very compleat; they
" returned ruined, worn out, and loft
" a vaft number of officers and foldiers;
" neverthelefs we had no confiderable
" action: the only one which was of
" any confequence, was to our advan-
" tage,

" tage. It was not by any vifible
" ftroke, but infenfibly, that our army
" wafted away.

" In effect, the greateft part of the
" detachments fent out to fkirmifh, and
" pofts that were diftant, and efcorts
" that were attacked by the enemy,
" were either furprized or beaten for
" want of difcipline in the foldiery, or
" by the neglect of the officers. The
" man is not yet born that ever faw an
" efcort march in good order: the
" foldiers continually employed in pil-
" laging, and ftealing away out of fight
" of the officer, get the habit of ftrag-
" gling from the beginning of the
" march, and you'll fcarce ever find an
" officer who gives the leaft attention to
" this abufe. It is the fame in pofts,
" parties, and detachments; either the
" foldier ftraggles, or if he ftays with
" his party, it is only to march in bad
" order, ftopping every minute, talk-
" ing when he ought to be filent, or
" murmuring when he ought to obey.
" If the enemy appears, they are ftupi-
" fied, and comprehend nothing, nei-

L 2   " ther

" ther do they know how to defend
" themfelves, or how to form ; there
" is nothing but confufion ; and if by
" accident any orders are given, which
" is feldom done, you are to fpeak to
" men who are deaf and immoveable.
" Being little ufed to military exercifes,
" or fubmiffion, or to the obedience
" due to their officers, they throw
" away their fire in the air, and are fure
" to be beaten ; and that, becaufe the
" foldier is not ufed to obey, and be-
" caufe we are never prompt enough
" in punifhing ; but efpecially, be-
" caufe young officers neither know
" how to command, nor make them-
" felves obeyed ; and thofe that do, often
" dare not, left they fhould incur the
" hatred of their companions, who ima-
" gine that punifhment makes foldiers
" defert."

Such was the opinion of one of the
greateft Captains France ever had ; an
opinion founded on experience and compleat knowledge, and which may furnifh
the beft leffons to officers who will reflect thereon. The decay of difcipline

was

was at all times the lofs of foldiers, and the caufe of fhame to officers, who difhonour themfelves lefs by the defect of courage, than by their neglecting to apply themfelves to their profeffion.

" You muft not imagine," fays this fame General, " that fubordination and " fervile obedience leffen a man's cou- " rage : it has been always feen, that " the fevereft difcipline has been efta- " blifhed, where the greateft exploits " have been done *."

Therefore an officer, who marches at the head of a body, ought to keep up the moft exact order, and a profound filence, fo that they may be always pre- pared to execute whatever movements he orders for their defence, however dangerous they may be. But in giving thefe orders, he ought always to appear confident and determined, which makes the foldiers think he is fure of his aim, and that he has taken the beft mea- fures.

Soldiers, when they fee their chief wavering and doubtful in his com- mands,

* Chap. I. of Difcipline, *Reveries.*

mands, imagine him to be at his wit's
end; and seeing him disturbed, they
themselves will be affected in the same
manner.

It is on these occasions, that an officer ought to keep his head clear, to
be able to manage his party, and to
be obeyed directly. The danger is
much greater on a march, than in an
attack; in the last, the soldiers have
their arms in their hands, and seeing
the enemy near, they are always ready
to engage; the contrary is observed on
a march, they are less on their guard,
and have not, in like manner, their
arms in their hands. "At such a
"time," says Vegetius *, "an attack
"stuns them, an ambuscade confounds
"them." An officer, who has a mind
to put himself out of danger of these
surprises, ought, when he gets at a little
distance from the camp, to make a
corporal or serjeant march eighteen or
twenty paces before him, with four or
five men, more or less, according to the
strength of his detachment: and also
two

* Lib III. Chap. II. of Institutions.

two or three at an equal distance on his
flanks, to make discoveries, and to examine hollow ways, swamps, lakes, or
ditches, which are on the right and left
of the road; to search into farm-houses,
barns, mills, and other places where the
enemy might lie in ambush. He ought
also to stop all peasants, men or women,
who are going the same road, and endeavour to pass him; and should march
them with him, till he is past all danger. I could mention several officers
of different ranks, who were surprised,
beat, or carried off, by neglecting some
of these precautions; but I will confine
myself to this example, which suits well
to my purpose *.

During the war in Spain in 1674,
Marshal Schomberg, who commanded
the French army, having a mind to
cover Roussillon, ordered a considerable
detachment to march, to secure the con-
voys that were coming from Perpignan

to

* The late war in Flanders, and the present in
America, have furnished some scandalous, and
some very unhappy instances, of this want of caution in officers, who shewed thereby, that their
rank was much higher than their judgment

to the village of St. John de Payés,
about three leagues from Perpignan.

This corps was pofted on a height
which was near- the high road, from
whence the commanding officer fent his
lieutenant and thirty men to take poffeffion of a chapel that was on an eminence ftill higher, at about three hundred paces from his poft ; from which
eminence the lieutenant could eafily difcover the Spanifh incampment on the
plain of Boulou, and over which their
parties muft pafs to attempt our con-
voys.

From Boulou to thefe two pofts there
was a long hollow way, through which
the enemy might march under cover;
and as furprizes were to be feared every
day, there was a detachment pofted alfo
in a cottage called the *red boufe*, with
orders to light fires to give notice to
the other guards, if they made any difcoveries, and to be always ready to help
one another.

A Spanifh officer, with forty horfe,
paffing the hollow way under favour of
the night, and being well acquainted
with

with the country, and the pofition of
the guards, lay in ambufh in the middle of the three pofts, with a defign to
furprize the lieutenant's detachment,
who went every morning to relieve the
guard of the chapel. This lieutenant
having got into the hollow way, thro'
which he was neceffarily to pafs, the
enemy fell upon him, and charged him
fo roughly, that all his men were either
killed or wounded, before he had time
to recollect himfelf. He received, for
his own part, two cuts of a broad fword
on the head, from the Spanifh officer,
who added to this treatment thefe in-
fulting words : " Go, fays he, learn an-
" other time to do your duty better,
" and to reconnoitre a place where you
" are to pafs with your guard."

I will not comment on this paffage,
which is taken from the relation of the
Catalonian war; becaufe I believe it
will be fufficient to read it once, to
prove what I faid, that it is neceffary to
examine every place where the enemy
may lie in ambufh to furprize you.
But as it is difficult, or rather impoffi-

ble, for a detachment that marches in a fuspected country, to examine all the villages by which they muft pafs, and where the inhabitants are oftentimes as much to be dreaded as the enemy; I think if an officer can avoid them, he had beft turn off a little, and come in again to the road when he has paffed them.

It is obvious, that for making thefe difcoveries, none fhould be employed but the oldeft foldiers of the party, whom you muft order never to ftop to drink, to divert themfelves or talk with the peafants, and never to lofe fight of the detachment; but to ftop all perfons that endeavour to pafs before them, and to come immediately to give an account of what they faw or heard to the commandant.

But as all the precautions that I have mentioned, do not remove the poffibility of an officer's being attacked upon his march; he muft, as foon as he fees the enemy, examine whether their party be greater than his own; whether it is horfe or foot, or both together. If they
are

are cavalry, and superior to him in
number, he is not to be discouraged
on that account; but, on the contrary,
he ought to avail himself of his own
advantages, by throwing himself into a
close country, uneven, or cut, which
may be difficult or inaccessible to the
horse. He must also raise the spirits of
his men by resolute and bold expres-
sions, and endeavour to make himself
master of some post, where he may be
able to maintain himself, while he sends
a faithful soldier to inform the General
of his situation. If in this situation the
enemy marches towards him, he must
do his endeavours to support the efforts
of their attack, ordering his men not
to be in a hurry, but to save their shot,
and not to fire till they can reach the
enemy with their bayonets. However
contemptible natural fortifications may
appear, such are found in every country by chance, which courageous men
have defended with extraordinary valour. The last age shews what seven
soldiers could do in one of these situa-
tions: The Duke of Rohan says, in his

Memoirs,

Memoirs, that they ſtop'd, for two whole days, before a poor houſe built of clay near Carlat, a whole army, which Marſhal Themines was leading to the county of Foix, conſiſting of ſeven thouſand foot, and five hundred horſe. If the road, in which a detachment is attacked on its march, be covered on either ſide with vines, wood, rocks, or by ſuch rough broken ground as may prevent the cavalry from penetrating it ; an officer, as I ſaid, ought to throw himſelf into it directly, and to continue his march towards his poſt by that way, keeping his men cloſe together, and always ready to receive the enemy.

If, on the other hand, the party of horſe which he perceives, be pretty near equal in ſtrength to his own detachment, he is not to diſcontinue his march on that account ; but ſhould form his men into a cloſe platoon of five files of ſix men each, if he has thirty ; of ſeven files of eight, if he has about fifty ; or of ten files of ten, if he has one hundred ; and thus, with their bayonets fixed,

fixed, prefenting their arms on every
fide, he is to continue to march towards his poft. An officer, who marches
in this manner, without lofing his order, and in filence, will convince the
enemy, that he is not afraid they fhould
fall upon him. But, however, if they
fhould, he muft halt his men, make his
firft rank kneel, pointing their bayonets
to the horfes breafts; the fecond muft
kneel alfo, prefenting their arms; and
the third fhall take aim over their heads.
You muft obferve here, that I only
fpeak of a detachment of thirty men;
for if the body is greater, you may
make two ranks aim at the fame time.
in this cafe, an officer muft forbid his
men to fire till the enemy's horfe are
within ten paces of the bayonets of the
firft rank; then the ftanding rank or
ranks which took aim, or were prefented, are to give fire, and to reload
immediately; thofe of the fecond rank
are to ftand up at the fame inftant, and
prefent, in order to fire, if the officer
commands them; but if the firft or
fecond fire has difconcerted the enemy,
he

he muſt order his men to riſe, and continue his march; always ready to begin
again, if the enemy ſhould return.

But if the enemy's party diſcovered
be ſuperior, conſiſting both of horſe and
foot, or of foot only; the officer muſt
endeavour to make himſelf maſter of a
mill, or a ſingle farm-houſe, to defend
himſelf till his General, to whom he has
given notice, ſends to diſengage him.
If he ſees no way to poſſeſs himſelf of
an advantageous poſt, or get to the
place he is detached for, he can do nothing better, than to fight his way re-
treating, and return to the camp, in
coaſting along a river, or wood *, if he
can,

* The tranſlator thinks the wood is preferable
to a river, eſpecially if it be cavalry that oppoſes
them, and that they can get that way to their
poſt or camp, and would recommend the following method, partly copied from that practiſed by
the Indians in North America; which is, as ſoon
as they get near the wood to break entirely, and
ruſh in with their firelocks in their hands, letting
them ſwing at arms length, ſo as to avoid the
branches of trees; and when they are got into the
thick part ſeven or eight yards, they are to halt:
it muſt be obſerved, that they will probably get
in at ſeveral diſtances from each other, and ſo
much

can, to avoid being furrounded; and if he
is fo clofely purfued, that he cannot avoid
being beaten and taken ; I fee no better
expedient to be adopted in this cafe,
than that of the *Barbetts* * of the vallies
of

much the better, as they can, by that means,
form the fooner, *in their manner*, which is in files
of five or fix men, or more, according to their
ftrength, leaving a fpace between every file, of
two feet or more ; fo that when they find the
enemy near them, and good cover for themfelves,
they face about, and form, as the place will allow them, either in one or more ranks, or they
advance by fingle files, to fire on the enemy, endeavouring always to cover themfelves behind
trees, or ftones, if they can ; and they may advance unperceived, by creeping on their bellies,
and by this means the enemy will be often de-
terred from purfuit, not knowing where they may
meet refiftance, or what to fire at; and the beft
way to deceive them is, when you are fairly in,
for the whole to lie down, and creep, as the officer fhall direct ; and if the enemy perfift in pur-
fuing you, after making your beft defence, you
muft retreat in the order of open files, ftopping
every now and then to amufe them with a fire,
which will greatly annoy and delay them

* *Barbetts* are peafants fubject to the King of
Sardinia, who abandon their dwellings when the
enemy has taken poffeffion of them The King
forms them into bodies, who defend the Alps,
being part of his dominions.

of Piedmont, who difperfe themfelves, and retiring from tree to tree, or from rock to rock, fo harrafs their purfuers, that they can neither beat them, nor take one man.

I promifed to mention fome remarkably clever *manœuvres* of private officers, which may ferve as proofs and explanations of the articles that I treat of. Therefore I cannot pafs over a march of M. de Beuvrigny, captain in the regiment of Cambrefis, which would do honour to a general officer : I take it from the hiftory of the revolutions of Genoa.

During the Corfican war in 1737 and 1738, the king fent reinforcements to the ifland, to reduce the malecontents to reafon. A convoy, efcorted by a frigate and two armed barks, appeared in the beginning of 1739, fteering towards San-Fiorenzo ; but they met with a dreadful ftorm the 8th of January, which difperfed them ; neverthelefs, all the veffels of this convoy arrived with four French battalions, at different ports of the ifland, except two tartanes, which

had

had the misfortune to run on shore the
same day, on the coast of the province of
Balagna, to the left of the river Oftre-
gone. M. de Beuvrigny, who commanded six companies of the regiment
of Cambresis, which were embarked in
these tartanes, saved himself and the
troops, by his presence of mind and resolution.

It was ten o'clock at night, when
the tartane, that this officer was on
board of, struck against the rocks,
with a dreadful shock, about one hundred paces from the coast: he hindered
his people from leaping into the sea,
where they must inevitably have perished;
and with a pistol in his hand, he com-
pelled the sailors to launch their boat,
and did not save himself, till after em-
barking successively all the sailors and
soldiers, which took up above two
hours.

He had no sooner got on shore with
his three companies, than he had intelligence brought him, with advice to
think of his safety; for if he stayed till
day-light, he run a risque of being at-

N   tacked

tacked by the Corficans: but he would
not abandon the three companies that
were in the other tartane, which was
alfo ftranded on a fand bank, at a little
diftance from the firft; the boat of
which was loft, in carrying fome of the
officers and foldiers on fhore, whofe
bodies M. de Beuvrigny knew on the
ftrand. He determined to affift thofe
that were ftill in the veffel, and made
his men go into fome cottages to warm
and reft themfelves for the remainder
of the night. At day break he fent
the boat to land his comrades, who
brought on fhore about fixty firelocks,
and a hundred and fixty charges of ammunition; half the mufkets were without locks, being taken off to prevent accidents on board.

M. de Beuvrigny having reviewed his
men, who amounted to one hundred
and forty only, pofted the foldiers,
without arms, in the middle; on the
flanks, thofe who had mufkets, without
locks, but with bayonets fixed; and in
the front and rear, thofe who had pieces
in order to fire. After this prudent difpofition,

pofition, he fet out for San-Fiorenzo,
from which he was about five leagues
diftant; but he foon had the Corficans
at his heels, who had heard of the ſhips
wrecked on their coaſt. M. de Beuvrigny croffed the river d'Oftrigone, in
their fight, having the water up to his
middle, and continued his route by a
mountain, in fpight of their ſhot, which
he anfwered now and then. He killed a
good many of the Corficans, and had
fome of his own men wounded; but in
fpight of the good care he took of his
ammunition, it was foon fpent, fo that
he had but five cartridges remaining
among the whole party; and had hardly
got half way, when a great body of Corficans appeared, of horfe and foot, pre-
paring to furround him, and to put all
to the fword: night approaching, his
men overcome with fatigue, without
guides, and without ammunition, feeing no other remedy, he determined to
furrender himfelf prifoner. The French
General greatly commended the bravery
of this officer; who being reclaimed in

the

the King's name, was immediately set
at liberty, with all his men.

The conduct of M. de Beuvrigny
was so prudent, and so well concerted,
that though it was not successful in the
end, I thought it my duty to mention
the whole ; his presence of mind in the
shipwreck, his zeal to save his soldiers,
his good dispositions in his retreat to
San-Fiorenzo. He withstood the repeated
attacks of the Corsicans for a long time,
and would certainly have retreated in
good order to that place, if his ammunition had not failed, and if he had not had
cold, hunger, thirst and fatigue, and a
rebel army in a dark night, to contend
with, in the midst of a revolted and un-
known country.

C H A P. VI.

Of the establishment of a body in a post.

THE moment of the establishing a
body in a post, is the most critical a detachment can be in. We have
often

often feen officers attacked at the inftant
when they thought they had nothing to
do, but to take, at their leifure, the
proper meafures to remain with fecurity
in the poft they had taken poffeffion of.

If the body that arrives at a poft is to
relieve another, the officer to be relieved is to put his guard under arms the
moment the centinels have informed
him that a new detachment is in fight.
This detachment being examined, he
may let them enter, in order to take
poffeffion of the poft, in the room of
thofe that are going out. The corporals are to go immediately and relieve
the centinels, and the officers and ferjeants are to give each other the neceffary orders to be obferved in the poft,
both for day and night: fometimes thofe
orders are given in writing, but often are
verbal. An officer, who commands in
a poft, fhould take. the greateft care to
remember them. He muft alfo en-
quire of the officer he relieves, whether
the enemy make incurfions into his
neighbourhood ? Whether their parties
are far off? Horfe or foot? And their
fituation.

fituation? After having procured the
exacteft information of thefe things, he
may take the beft precautions to put his
poft out of danger of furprize.

As foon as the corporals are returned
from relieving the centries, the officer
who is going off, is to form his party
into ranks, and march back to camp,
with the fame precautions that he fet
out. The new guard are to remain
under arms until the old one has got off
about twenty or thirty paces: then if
they are pofted in a redoubt, the commander muft make his men lay their
arms on the top of the parapet, and fee
that they cover the locks with their
haver-facks, in order to preferve both
them and the priming from duft, dirt,
or moifture.*

But if an officer has relieved a party
in the open field, in a poft without any
fortifications, and where he does not
chufe to make any, he muft order his
men

* However it feems better, inftead of laying
the mufkets, as our author directs, on the top of
the parapet, to reft them on the *banquette*, leaning
their barrels againft the parapet.

[95]

men to ground their arms in the day
time, and not to ftir far from them;
and to keep them between their knees
at night, as they fit round their fire,*
obferving to turn the locks inwards, for
fear of accidents. Thefe precautions
being taken, the officer is to go vifit the
centries, and the places round his poft,
to know where to retreat to, in cafe of
an attack.

But if an officer is detached to a poft
that has not been occupied before, as
foon as he arrives, he muft poft his centries to prevent his being furprized from
without,

* The tranflator thinks thefe fires very dangerous for a party in an open country, without a redoubt; unlefs you poft centries at a confiderable
diftance from the fire, round about you, to obferve the approach of an enemy; for thofe at the
fire can fee nothing at a diftance, fo that the
enemy may get between them and their camp, or
place of retreat, unperceived, even till they are
within twelve or fourteen yards of the fire; nor
can they be heard, as the burning of the wood
and wind together, make a loud noife; and when
got fo clofe, a very fmall party will (by boldly
running in) with bayonets fixed, carry off a much
larger party fitting down, and who have not half
a minute to rife, and too confufed to think.

without, and place his arms as I directed before. But if he is sent to a
mill, a house, or cottage, he must draw
up his men in order of battle about fifteen or twenty paces from the place,
and send a serjeant or corporal, with
five or six men, to examine the cellars,
the chambers, and garrets ; which being
done, he may post his centries, and take
possession of the place ; and make every
soldier place his arms so, that each may
find his own without confusion : He
must lodge the inhabitants elsewhere,
and intrench himself in the manner before-mentioned for single houses.

Lastly, if an officer is to establish
himself in a village, as it would be very
difficult for him to examine every house
and place where the enemy might lie
concealed, he should send to the magis-
trate and people of the first rank, keeping his men drawn up under arms, at
about fifteen paces from the village,
and oblige them to declare, in the
King's name, whether there are any
hostile parties, suspected people, or
concealed arms, in the place ; after
which

which, he is to send to poft his centries, and enter the village, and poft fmall detachments at the avenues of five or fix men, according to his ftrength ; then examine the *chateau* and church, or any other building that ftands by itfelf, for fixing his chief poft, and the place of his laft retreat, if he fhould be forced in his advanced pofts.

After an officer has taken poffeffion of a poft, he fhould go and fee how the corporals have pofted the centries ; if he finds they are improperly placed, he is to change them. When many centries are to be pofted, care fhould be taken that the oldeft foldiers, or thofe that know their duty beft, fhould be pofted at the moft expofed and dif-tant places, and in fuch a manner, that they may difcover all approaches to the poft ; fometimes he may place a few in trees, fo that they may fee a far off, and not be feen themfelves by the enemy.

After the officer has made this round, he is to vifit the places about his poft, to fee if it be neceffary to cut up any
roads,

roads, or to ſtop them by *abbatis*, to pitch on the places where trenches may be dug, and what ſort of intrenchment he ought to make to ſecure himſelf, and how he may avail himſelf of all the little contrivances before-mentioned. If there ſhould be near at hand a hollow way, or a thicket of wood, a houſe, or any other covered place, which the enemy may take poſſeſſion of, and where they may lie in ambuſh, in order to fall afterwards on the poſt; here he ſhould place a ſmall guard of ſix or ſeven men, commanded by a ſerjeant or corporal, who are to have orders to fall back to the chief poſt if they ſhould be attacked, or to ſupport themſelves till they can receive aſſiſtance. The men of this little guard muſt be charged to make no fires, becauſe the light, or ſmoke, would be a guide to the enemy, who would chuſe to avoid it, if they intended to ſurprize the principal poſt. Experienced officers and ſoldiers light fires in places where they have no guards, to make the enemy imagine they have poſts every where, and make

their

their ambufcades in places where they
have no fires. This *fineſſe* may alſo be
uſed in all poſts that lie in an open
country, from which the officer, to
this end, muſt detach two or three men
to go, during the night, from one fire
to another to keep them up.

After this external arrangement, and
when the centries are poſted at the avenues, on the bridges, and the tops of
the ſteeples, the officer is to ſee what
ſort of intrenchment is beſt for the defence of the poſt; and as ſoon as he is
determined in this point, he is to mark
them out, and make the workmen finiſh
them; then he is to place therein the
little guards, which are appointed for
their defence; and if the detachment is
to continue for many days, he is to
lodge the men in the neareſt houſes,
giving order to the magiſtrate to furniſh
them with ſtraw. Care muſt alſo be taken, that the ſoldiers ſo lodged be not
ſuffered to ſtray from the houſes, that
they always lay their arms where they
can find them eaſily, and, if poſſible,
there ought to be an officer to command

at

at every one of thofe pofts, or a ferjeant
or corporal at leaft; the foldiers who
guard the intrenchment or fort, are to
fit all night on the *banquette*, leaning
their mufkets on or againft the parapet,
and to take care to be alert all the time.
Let the enemy be at what diftance they
will, the officer ought never to fleep but
with his cloaths on, fo that he may be
always ready to go wherever his pre-
fence may be wanting; and to make
his ferjeants and corporals go the rounds
often to vifit the centries *.

If he has chofen the *chateau* or *man-
fion-houfe*, the church, or parfonage, or
any other houfe, for the chief poft of the
village, and that thefe places be inha-
bited, the inhabitants muft be lodged
elfewhere, fo that no body fhould ftay
in the poft to hinder or betray him, or
any

* As nothing can be more injurious to the character of an officer than being furprized in his
poft, and as ferjeants or corporals are not always to be depended on, the tranflator would
recommend it to the officer to go the rounds himfelf two or three times a night, without mentioni-
ing his hours; and to order the ferjeants or corporals to go at leaft every hour.

any way obſtruct him in carrying on
whatever works he may think neceſſary. An officer ſhould never ſuffer himſelf to think, that the little time
a detachment is to ſtay at a poſt, ſhould
be a reaſon to neglect preparations for
its defence. " When an officer is in a
" poſt," ſays M. Vauban *, " he ſhould
" intrench himſelf immediately, tho'
" he were to ſtay but four hours."

To which I will add, that all his
works ought to be well made, and ſo
contrived, as to defend every ſide where
the enemy may come.

Monſieur Follard gives us on this
ſubject an excellent maxim, which may
ſerve as a general rule. " You muſt,"
ſays this author †, " *imagine* that your
" poſt is attacked, in order to *imagine*
" the method of defending it." And
M. le Baron de Travers, in his Obſer-
vations on the Art of War by Monſ.
de Puſſegur, ſays ‡, " that poſts ought
" always

---

* Attack and Defence of Places, Tom. II.
page 180.
‡ Tom. V. Defence of Poſts.
† Chap. 10. ſecond part.

" always to have strength and means
" of defence, in proportion to what the
" enemy may employ against them."

M. Duclaux de Barrieres, captain in
the regiment of Lorrain, beyond any
one I ever saw, seemed persuaded of
the utility of speedily intrenching his
party; when this officer was to stay a
few hours in a post, he immediately
made an *abbatis* of trees; and if he was
in a village, he directly intrenched the
chateau or principal house.

C H A P. VII.

*Of precautions to be taken in a post, to
avoid a surprize.*

I HAVE said, that the security of an
army depends on the vigilance of
its advanced guards : And is it not a
matter of the highest importance to a
General, supposing him to have made
the best disposition possible, that the
officers detached from his army should
know

know how to go about, and be fure to
execute, whatever was committed to
their charge ? " The chief object that
" a foldier ought to have in view, when
" he is detached," fays Monfieur de
Vauban, in his Treatife of War, " is
" always to forefee any bad accident
" that may happen to him." The
neglect of regularity, or the leaft re-
laxation in duty on a poft, may have
very unhappy confequences. Hiftory
furnifhes us with many examples, where
camps have been furprized, and armies
cut in pieces, by the neglect of detachments, who ought to have watched for
their prefervation.

After an officer is eftablifhed in his
poft, his principal concern fhould be,
how to provide againft an unexpected
attack, his being betrayed, or carried
off.

M. de Travers fays *, " the only
" means in war, to be fafe from fur-
" prizes, confift in taking precautions
" againft every thing the enemy can
" poffibly undertake ; therefore you
" are

* Supplement to Military Study.

" are not to reckon that your fafety
" depends on the probable, but on the
" poffible diftance of the enemy."

To avoid the ill confequences that
may refult from negleƈt, an officer,
when he is fent to a redoubt, or any
other detached poft, fhould let no ftrangers whatever enter into it, not even
any foldier that did not belong to his
detachment: he fhould forbid all his
own people to pafs the bounds * where
the centries are fixed, or to ftraggle
from the poft, under any pretence what-
foever; and he fhould call over the roll
three or four times a day.

When the corporal is going to relieve the centries, the officer is to examine the foldiers who are to relieve
them, before they march off: he is to
vifit them after they are pofted, and to
make his ferjeants and corporals vifit
them now and then.

When it grows dark, he is to make
the centries come nearer to his poft, fo
that by forming a lefs circle, nothing

may

* By order of the 1ft of July 1727, every foldiers, who paffes the limits, fhall be hanged.

may pafs unperceived between two of
them. If there are fignals to be made,
or to anfwer, they muft be ordered to
be very attentive. If, in the neighbourhood of the poft, fome parts are
much wooded, in fuch places two centries ought to be pofted, with ftrict orders not to fpeak to each other, or walk
about; and fome of them may, as I
faid before, be pofted in trees, minding
to relieve them every two hours, or every
hour, if the weather is fevere. If an officer, in making the round of his centinels, finds any new foldiers among
them, he fhould remind them of the
duty they are bound to obferve while
they ftand centry, and tell them they
ought never to quit their poft, nor fall
afleep upon it; and that they are not
to fuffer themfelves to be relieved by
any but their own corporals; and to
permit no foldier to go from their poft,
and to inform their officers of any thing
they fee; to ftop any body that may
advance towards them, till they know
them, and to fire on thofe who refufe
to anfwer, after calling out three times,

P   *who*

who comes there? laftly, if they find them
ftill approaching, they are to return to
the poft of their commanding officer.

It is not fufficient for an officer, who
commands in a poft, merely to receive
reports from his centinels ; but he
fhould, as it were, for paftime, fhew
his detachment the way to defend themfelves, if they fhould be attacked ; and
explain to them, if the enemy made
fuch an attempt, how he would oppofe
them by another ; if they undertook
this, he would ufe that, and baffle them
in every point. He may make fome of
his men try to fcale the works, to fhew
all his people the difficulty of executing
it. By exercifing them in this manner,
he will prepare them to refift the enemy
with eafe : thus he will give them a
high opinion of themfelves, and make
them put great confidence in him. He
muft avoid all this time, while he treats
them as comrades, not to be too fami-
liar with them; for in that cafe, if in
the midft of an attack he fhould order
them to do fomething that was not to
their liking, inftead of obedience, they
might

might perhaps mutiny, and difobey his orders.

After the taking of Bellegarde in Rouffillon, by Marfhal Schomberg, in 1675, there happened an event, which fhews the importance of what I now fay : I take it from the relation of the wars of Catalonia; I will give the whole, being analogous to my fubject.

At a league from this fortrefs, on the road to Colioure, there was a chapel dedicated to the Virgin: this chapel was built in the middle of many rocks, whofe points, almoft inacceffible, ferved it for a wall, fo that it was very ftrong, as well by fituation, as from the rocks being proof againft cannon-fhot.

Marfhal Schomberg, being refolved to make himfelf mafter of this important poft, which was defended by a Spanifh captain and fifty Germans, detached a large body from the army, commanded by M. de Gaffion, marfhal de camp : the trenches were opened, fome cannon were drawn thither by men, and a battery was erected on a rock near the poft ; but it had little

effect,

effect, though it was well ſerved. The captain and his ſoldiers laughed at the beſiegers for five days, and ſeemed likely to be in the ſame condition for a long time to come ; but a cannon-ſhot having killed three Germans, who were looking over the wall, all the reſt loſt their courage, and told their captain, in an inſolent manner, to ſubmit, and make his compoſition, or they would do it for him ; he was aſtoniſhed at their cowardice, remonſtrated to them their duty, but in vain ; they hung out the white flag. The French, delighted with their diſobedience, conſented to treat with them ; but while the Germans were all in confuſion, the French got poſſeſſion of the chapel-gate, before any conditions could be drawn up, and they were all made priſoners.

This example, which ſhews how difficult it is to force a brave man in his poſt; and how important it is for a commander to maintain an awful authority over his own ſoldiers, ſhews alſo that one ought never to deſpair, on account of the revolt of a few mutineers. The
ſpirit

spirit of rebellion never seizes a whole
body at the same inftant; but it is in-
finuated, and fpread by the feditious
propofitions * of a few. As foon as an
officer perceives this, he fhould command immediate filence; and if they
have the infolence to continue the tu-
mult, he fhould take inftantly a mufket,
fhoot the moft daring of the mutineers,
and threaten to hang all thofe that fhall
refufe obedience. I could account for
the reafonablenefs of this conduct, and
relate many examples to prove, that
this way, though it may appear violent,
is the only one to check a mutiny in
foldiers, or even in a mob. But as that
would be foreign to the work I am
upon, I return to my fubject.

When an officer has fhewn his men
the advantage that an intrenched body
has over thofe that attack them openly,
he muft take care to keep up good
order, and not to fuffer himfelf to be
enfnared by the enemy.

If

* By order, of the 1ft of July, 1727, mutinous
foldiers are to be fent to the provoft to be hanged
and ftrangled.

If an officer is detached towards the limits, and that deferters come to his poft, he muft be cautious not to let them in. He ought, on the contrary, to make them remain without on the glacis, and fend two or three foldiers to receive their arms, and conduct them to the General of the army, efcorted by fome mufketeers. If thefe deferters are very numerous, and that he cannot fpare men enough to conduct them fafely, without weakening his own poft too much, he muft write to the Adjutant-General, to beg of him to fend a detachment to receive them.*

It

* If thofe deferters fhould offer themfelves at an advanced poft, near or on the glacis, in the time of a fiege, and are too ftrong to be received in the poft; if the above method is followed to fend men to receive their arms, or to keep them there till the General can fend to receive them, the whole will, in all probability, be knocked in the head from the couvert way, or be retaken. Therefore, if they were ordered, on their approach to the poft, to throw down their arms; when that's done, the officer may fend as many armed men, as he can fpare, to conduct them to the trenches.

It is not alone fufficient to take the fteps that I have fhewn for the internal fecurity of a poft that is to be defended. But a fkilful officer ought alfo to look to the external fafety, and endeavour to difcover the defigns that the enemy may form againft him. The moft critical time for detached officers to be alert, is an hour or two before day ; great care muft be taken to keep the foldiers awake, and to make them fit on the *banquette,* each clofe to his mufket. One or two patroles ought to be fent out, during the night, and at day break, as fcouts, to make difcoveries in the environs. Thofe patroles fhould be of four or five men, and be ordered to march flowly, and with the leaft noife poffible, to examine the hollow ways, hedges, ditches, woods, and neighbouring houfes, to ftop and hearken every now and then, to hear if there be any noife, and to return in half an hour, fo that another patrole may go out immediately after them.

It fometimes happens, when two armies are encamped oppofite each other,

and

and that there are many posts in one
and the same line, that two night patroles meet each other. Then, as it is
not possible to discern whether they are
friends or foes, that patrole, which has
first seen the other, should slip on one
side of the road and lie down behind
some bushes, or in a ditch, so as to ob-
serve their approach, and examine if the
other be the strongest: if so, they are
to let them pass, without saying a word,
and then to return to their post, to give
intelligence of what they saw. But if,
on the contrary, it should be weaker,
the officer should make the signal that
was given in orders, or that which his
officer gave him for the patroles of the
night. This signal is commonly to strike
one or two strokes on their pouch, or
on the stock of their musket; to which
the other is to answer by an agreed
number according to order. If the
other patrole will not answer, the first is
to march up to them, with bayonets
fixed, and fire on them, if they endeavour to retire, and oblige them to lay
down their arms. I saw, during the
war

war in Italy in 1745, old foldiers, who of their own accord, afked leave to go out fcouting, and took a great deal of pleafure in it.

When a detachment is to be pofted oppofite the enemy, one may expect to be attacked; therefore they fhould advance fome little parties during the night, about twenty-five or thirty paces from the poft, who are to lie down on their bellies, between the centries, at the places where they think the enemy may come: and orders fhould be given to the commanders of thofe little bodies, to make a fingle foldier reconnoitre any parties they may fee; fo that they may not miftake the friendly patroles for parties of the enemy; and all fhould return to their chief poft on the firft report of a mufket.

When one is charged with the defence of a poft that ftands fingle, it is of the greateft confequence not to neglect any of the precautions that I have before mentioned; but there are others as effential to be known, when a village or hamlet is to be defended. An

Q   officer

officer fent to a poft of this kind, fhould be careful to prevent any fufpected perfons from flipping in, and to keep the peafants from revolting. To this end he fhould order, by means of the magiftrate, two of the moft noted peafants in the place, to be pofted with the centries on duty, at the two only entrances into the place, all others being ftopp'd up by the entrenchment. Thofe peafants are to be relieved every two hours, and are to be ordered to examine the inhabitants that may come in or go out of the village ; and the whole of the inhabitants fhall be declared refponfible for any accidents that may happen thro' the treachery of thefe peafant centinels, or if thro' their neglect any of the enemy fhould get into the village difguifed.

It fhould be given in orders to the foldiers who guard the intrenchments, not to fuffer any peafants to approach them ; and to thofe who are pofted at the paffes, to ftop them up at night, by laying two trees acrofs, and not to open them till next day ; and alfo to fearch all waggons loaded with hay or ftraw, cafks,

[115]

cafks, or any thing elſe, by thruſting in
iron ſpits, or their ſwords, to feel that
there are no men, arms, or ammunition
concealed therein.

  During all the time that a detachment
occupies a village, the inhabitants muſt
not be allowed to hold fairs or markets,
or make proceſſions ; for often thoſe aſ-
ſemblies afford opportunity to the enemy,
to ſlip in and ſurprize the place. Poli-
bius gives us a leſſon on this ſubject,
which may not be found diſagreeable
here.

  " The unhappy conſequences of this
" liberty have been experienced above a
" hundred times," ſays the * tranſlator
of this author, " and yet no remedy is
" applied. How unjuſtly does man
" paſs for the moſt artful of all animals ?
" There are none more eaſily ſurprized.
" For how many camps, how many
" garriſons, and how many poſts have
" been loſt by this liberty ? This mis-
" fortune has happened to an infinite
" number, and nevertheleſs we are ſtill
" novices, as to this ſort of ſurprize."

Q 2   An

* Don Vincent Thuillier, t. 5. liv. 16.

An officer, who commands in a post
cannot be too watchful, lest there should
be any plots against his safety. The
enterprize of the enemies on Brisac, in
the month of November, in 1704,
comes too nigh my subject, to be passed
over in silence. The Governor of Fri-
bourg, having formed a design against
that place, set out at night on the 9th
or 10th of that month, with two thousand men, a great many waggons; some
of which were loaded with arms, grenades, firelocks and pitch, and the
others with chosen soldiers. All these
waggons were conducted by officers,
disguised like waggoners, and were covered with poles, with hay over them;
so that they seemed to be only loaded
with contribution hay. In this manner, and being favoured by a thick fog,
they arrived at eight o'clock in the morning, at the new gate. Two of the
waggons with men, and one with arms,
got into the town immediately: but an
Irishman, overseer for the undertakers
of the fortifications, seeing thirty men
near the gate, who had not the appearance

ance of peafants, though they had the
drefs, afked them who they were? and
why they did not go to work, like the
others? Upon their not anfwering, or
appearing confounded, he ftruck fome
of them with his cane; upon which, the
difguifed officers feized the arms that
were in the waggon next them, and fired
fifteen or twenty fhots at him, within
five or fix paces diftance, without wound-
ing him; he threw himfelf into the ditch
immediately, where they fired feveral
fhot more at him, to no purpofe, while
he cried out loudly, *to arms*. At this
noife, the advanced guard of the half
moon, and that of the gate, took arms,
and tried to hale up the bridges; but
they could not, on account of the waggons that the enemy had ftopped on
them. The officers and foldiers, who
were in the waggons, leaped down, and
took to arms; and joining the reft, attacked the guard commanded by a Captain of grenadiers, but they were repulfed; and five being killed, the reft
were difheartened, and fled, fome into
the town, and fome out: then he fhut
the

the firft gate, through which, being of
pallifadoes, or · rails, the enemy, who
were on the bridge, fired at every thing
that appeared. The Captain left there
one half of his guard, and mounting
the ramparts with the reft, kept a con-
tinual fire on the enemy. A Lieutenant, who commanded twelve men in an
advanced guard, was attacked at the
fame time, by an officer who clapp'd a
piftol to his breaft ; but he wrenched it
out of his hand, and fhot him dead.
This Lieutenant defended himfelf till
the end of the action ; but having received many wounds, he died the fame
day. On the noife of this furprize, M.
de Raouffet, commandant of the place,
diftributed the garrifon in all the neceffary pofts, and did every thing that
might be expected from a good and experienced officer. In fine, the enemy
feeing their defign fail, retired in diforder, leaving behind them a great
many waggons, and above forty foldiers,
who were either killed or wounded. Such
was the enterprize on Briffac, which
only failed by an effect of meer chance.

The

The excellent *manœuvre* of M. Vedel captain, and fince lieutenant-colonel, of the regiment of the Ifle of France, is a more recent example, to fhew of what importance it is to an officer, to ufe precautions when he is detached to a poft. During the troubles in the ifland of Corfica in 1739, this officer being detached to Chifoni, a village in that ifland, the parfon of the parifh afked the officer commanding in the open country, to permit the penitents of a neighbouring convent to come to this village, according, as he faid, to their annual cuftom, of making a proceffion to a certain chapel in the place, which he named. The commandant confented ; but M. Vedel, who was detached thither with fifty men of his regiment, being furprifed to fee fo numerous a proceffion coming into the village, which was compofed only of the peafants of a revolted country, called *to arms*, drew up his men, and thus difconcerted their projects ; in fact, many of thefe penitents, whom they feized, were found to be armed with

piftols

piftols and fwords; an account whereof
being fent to Marfhal Maillebois, then
General of the French troops in Corfica, he commended the activity of M.
Vedel, and ordered fome of the penitents, parfon and all, to be hanged
up on the fpot.

This example, and a thoufand others
that I could cite, fhew, that an officer
who commands in a poft, ought to be
very attentive, left he fhould fall into
the fnares the enemy may lay for him:
the lofs of a poft, though it may appear
very trifling, may have the worft of
confequences. The furprife of Amiens
in Picardy, in 1597, will ever be remembered. The Spaniards having
formed a defign to furprife this town,
laid fome foldiers, difguifed like peafants, in ambufh, in a houfe near the
gate, and fent forward a waggon loaded
with walnuts, of which the driver fpilt
a fack, as if by accident, and whilft the
foldiers of the guard ran crowding to
gather them up, the difguifed Spaniards
fell on them fword in hand, furprifed
the gate, and made themfelves mafters

of

of the town, which coft Henry IV. fix
months and a half fiege. Thefe events,
which are frequent in hiftory, fhew
how alert an officer ought always to be;
the many ways there are to furprife a
poft, ought to make one more miftruft-
ful, to be always ready to parry any
blow, and to make it abortive.

If any ftrangers or neighbouring peafants come to fee their friends or relations, the centries fhould ftop them, and
fend notice to the commandant, who is
not to let them in, till after the chief
man of the place, the parfon, the magiftrate, or fome of the moft confiderable inhabitants, promife to be anfwer-
able for them. Note alfo, that this per-
miffion fhould only be granted on working days, not on Sundays or holidays, as
all the peafants are idle on thofe days.
This precaution, relating to the enemy
without, is effential; but another, which
is no lefs fo, is to be on one's guard
againft the inhabitants of the village itfelf, who, in an enemy's country, are
always ready to betray, or to revolt.
The commandant of the detachment
R   ought

ought to take one or two of the magiftrate's children, or three or four from
the moft noted families, whom he is to
keep in his principal poft, as pledges
of the fidelity of the inhabitants: Great
care fhould be taken that thefe children
fhould receive no ill treatment; that
they fhould not be detained longer than
half a day each; and that they be relieved by others. The commandant
fhould forbid the inhabitants to gather
together in public-houfes, or the walks,
or any where elfe; and fhould have this
order pafted up over the church door.
If, after they come out of church, they
fhould ftop in the open places to talk
together, he fhould fend the patroles
to make them difperfe. Orders fhould
be fent to all the inn-keepers, and all
the inhabitants, not to receive any
ftranger into their houfes, without fending notice thereof to the commandant:
and they muft be ordered not to ftir
out of doors after the drum has beaten
the *retreat*, on pain of being fhot by
the centries who may fee them; or
feized and carried to the black-hole by
the

the patroles, who fhould march flowly,
and ftop now and then to hearken if
there be any noife; to go through all
the quarters that fhall be appointed for
them; and go to give an account to
the commandant of what they difcover,
that may occafion any alarm in the
poft.

If a quarter of the town fhould be
on fire, or if the inhabitants fhould
quarrel among themfelves, an officer
fhould then be very cautious how he
fends his guard to affift them, as thofe
are often fnares of the enemy to try to
divide the forces of a detachment, in
order to attack them afterwards. A
commandant, on the contrary, fhould
order the alarm-bell to be rung, and
make all the pofts, that defend the village, to take up arms; and order thofe
who command them, to make their
men ftand to the parapet with their
arms, to watch all that paffes without
fide the village. The foldiers of the
chief poft fhould alfo be under arms,
and the commandant fhould, at the
fame time, fend four or five men with

a ferjeant or corporal, to part the fray,
or make the inhabitants work to put
out the fire.

As all the precautions neceffary for
the fafety of a poft are too many to be
remembered, an officer fhould give his
orders in writing, and have them pafted
up, particularly at every little feparate
poft or guard.

Officers, on detachment in villages,
fhould efpecially be careful not to op-
prefs the inhabitants, by demanding ex-
orbitant fupplies. I know that it is
fometimes permitted, by an order of
the * General, to exact fire-wood, forage,
candles and oil, for the feveral guards ;
but thefe demands ought to be proportioned to the abilities of the inhabitants.
I could mention here many examples of
officers, who have, in a bafe manner,
abufed this power, by increafing, to
fuch a degree, thofe contributions, that
the magiftrates have been obliged to
pay them in money, not being able to
fupply

* By an order of the 30th of November 1710,
it was forbid to exact any thing in the villages,
without paying for it.

supply them in kind. An officer therefore cannot be too delicate on thefe occafions; and fhould fee that the inhabitants are not pillaged * or ill-treated
by the foldiers. Every thing is to be
dreaded from enraged people; and if
the lofs of our wealth makes us loofe
our fenfes, as is faid, to what defpair
will not people be drove, who feeing
their country ravaged, their effects pillaged, and laftly their perfons abufed,
and treated like flaves? I will not fay
that humanity calls aloud againft fuch
rigorous treatment, becaufe it is too
common to fee war filence the laws of
humanity; but I will fay, that not only
fmall detachments, but even whole garrifons, have been driven out, and had
their throats cut, in the towns they defended, by the inhabitants, whom they
had reduced to defpair.

History abounds with examples of
this kind; but without enumerating
them here, I will confine myfelf to that
which

---

* By an order of the 8th of April 1718, it was
forbid, on pain of death, to foldiers, to fteal any
thing, to fcale walls, or break open houfes, &c.

which the town of Genoa exhibited to
all Europe, at the end of the year
1745.

The Auſtrians having made themſelves maſters of this capital, the Marquis de Botta was appointed commandant, and had under his orders a large
garriſon of Germans, who treated the
Genoeſe with all the rigour imaginable;
while, by orders from the court, they
loaded them with contributions. This
General having reſolved to take away
ſome artillery that was on the ramparts,
on the fifth of December in the ſame
year, the bed of a mortar broke in a
narrow ſtreet; the populace gathered
about, but the officer, who inſpected
the removal, having ſtruck a Genoeſe
with his cane, who ſtood in the way,
or refuſed to help, the latter ſtabbed
him inſtantly in the belly with his knife.
The commotion becoming general, the
inhabitants flew to the arſenal, broke
open the gates, and took out the arms,
repulſed the Auſtrians from ſtreet to
ſtreet, and drove them out of their
town, after killing above five thouſand
of

of them. A good leſſon, on which I
intreat all military people to reflect.

CHAP. VIII.

*Of diſpoſitions neceſſary to maintain a party
in a poſt.*

BUT it is not ſufficient, for the pre-
ſervation of a poſt, to have made
good intrenchments, and to have taken
precautions againſt all kinds of ſurpriſes;
for as the enemy may attack it with
ſuperior forces, thoſe who are attacked
ſhould make their diſpoſitions ſo as not
to embarraſs one another, and that every
arm may be ſo properly placed, that all
may contribute to the common defence.

If it is a redoubt that is to be defended, or whatever other intrenchment of earth, ſeven or eight trees, with
all their branches, ſhould be reſerved,
to ſtop the breaches that the enemy
may make; the parapet ſhould be lined
with all the ſoldiers of the party, and
arm

aim the firſt and ſecond ranks with their
muſkets, and their bayonets fixed, who
are not to fire until the enemy are on
the glacis ; if poſſible, the third rank
ſhould be armed with long. weapons,
ſuch as ſpontoons, halberts, lances, forks,
or, as M. Follard ſays in his notes * on
Polybius, with long poles, having bayonets fixed at their ends. Theſe long
weapons will ſtop the enemy at the edge
of the ditch, or at the outer edge of the
parapet, where it will be eaſy to ſhoot
them. Theſe ſoldiers, of the third rank,
may alſo be furniſhed with grenades, or
faggots well lighted, to throw among
the enemy that have leaped into the
ditch ; alſo aſhes or ſlack lime may be
thrown on them, the burning duſt of
which will infallibly blind them. Tho'
this laſt expedient may ſeem extraordinary, I believe, after many trials, I
may take upon me to anſwer for its ſucceſs.

It is evident, that the different methods I have been ſpeaking of, to arm
ſoldiers for the defence of a parapet,
cannot

* Tom. III. page 278.

cannot be practised by a small body of
thirty or fifty men; this number not
being sufficient to form two or three
deep, they are to be armed with their
muskets and bayonets only; and if the
enemy gains the parapet, they must be
resisted sword in hand, keeping always
quite close to the parapet: care must
be also taken to post eight or ten soldiers,
more or less, according to your numbers, in the ditch, at the parts the least
exposed, and least in sight of the enemy;
to keep in this position till the enemy
leap into the ditch; then with orders to
divide into two parts, one to the right,
and the other to the left, to fall on their
flanks, with their bayonets fixed. This
kind of sally will astonish the enemy
greatly, as those who attack, don't
dream of being attacked; but, on the
contrary, are surprized to see themselves
so hotly charged.

The parapet of a redan, is to be lined
in the same way as that of a redoubt,
observing if the right or left of those
redans were joined to any heights, or
commanded by any rocks, which often

S   hap-

happens ; they fhould be taken poffeffion of by feven or eight foldiers, covered with an *abatis*, to hinder the
enemy from making themfelves mafters
of them ; and that they fhould not over-
whelm thofe in the intrenchments, by
throwing down heaps of ftones.

If it be a *chateau*, a houfe, a cottage,
or a mill fortified with a curving parapet,
that is to be defended ; a part of the
foldiers defigned for the defence of the
intrenchment, are to be pofted, as I
have juft now faid. This firft difpofition being made, it is not immediately
neceffary to place foldiers in the ground
floor at the loop holes ; as they will be
ufelefs there while the outwork can be
maintained : but if thofe who defend
it, are forced and obliged to abandon it,
they are to take refuge in the houfe, and
poft themfelves at the loop holes. Two
of the ftrongeft foldiers are, at the fame
time, to be placed at each jaum of the
door, withinfide, with their bayonets
fixed, to ftab the enemy, who fhall attempt to enter, and to pafs the defile
made

made by the tree placed before the
door.

An officer, who as I faid in the Chap.
Of precautions to be taken in a poft, will
acquaint his men before-hand, with the
different manœuvres that are to be performed, in cafe they fhould be attacked,
need not be afraid that they fhould execute, what I am now fpeaking of, in
diforder. The foldiers at the loop holes
fhould never fire, until they are fure of
their mark, and mind that one of them
keeps the muzzle of his piece always in
the loop hole, while the other is charg-
ing.

There fhould be alfo at the loop
holes, of the firft ftory, two or three
men to annoy the enemy, by mufket
fhot; and there fhould be a forked pole
left befide each loop-hole, of ten or
twelve feet long, to thruft occafionally
thro' the holes, to grapple and overturn
any ladders the enemy might lean againft
the walls, obferving to pufh them quick
and ftrong, fo as to overturn, at the
fame time, both the ladders and the
men who are on them.

If

[132]

If the windows of the first story are not quite stopped up, and though the floor is cut away before it, two soldiers may be posted near it, to overturn the enemy's ladders: Lastly, some soldiers should be sent up to the second story, which is generally the uppermost in peasants houses in the country; they are to be posted at the brink, the walls where the tiles were taken off, with orders to shower down stones, ashes, lime, or half burnt dung, on the besiegers, and to beat down their ladders with the rafters of the roof, in order to prevent them from gaining the top of the house.

If it is a village that is to be defended, and that little guards are posted at the entrance of the streets, it will be proper to shew each of them, in what manner they are to retreat; if being forced, they are obliged to fall back to the principal post, defending themselves from house to house, and from street to street, behind the trenches, that they have cut across them.

If there be a few cavalry in the detachment, they should be posted in the market

market place, or any open ſtreet, where
they may be ready to fall on the enemy
ſword in hand, as ſoon as they appear
expoſed; but if they are numerous,
they may do the ſervice of infantry with
ſucceſs.

Laſtly, If there are cannon, they
ſhould be placed oppoſite to the ſtreets
that lead to the chief poſt, to keep the
enemy at a diſtance.

When all theſe diſpoſitions are made,
an officer ought to order each and every
one of his ſoldiers, to remain at the
poſt aſſigned to them; to make a little
fire if the ſeaſon is cold, and to place
their arms ſo, that they may find them
readily, and without confuſion.

CHAP.

CHAP. IX.

The defence of posts.

THE obstinate defence of a post is an action wherein an officer may acquire the greatest glory : this resistance is not so much owing to the number of soldiers destined for its defence, as to the ability of the officer who commands ; it is in him chiefly that the strength of the intrenchment exists ; and if to his determined bravery the talents necessary on those occasions are also added, and that he knows how to persuade his soldiers, that the treatment they are to expect from the enemy is a thousand times worse than death, one may say, that he will, in some sort, render his post impregnable.

If an officer, posted in a redoubt, is attacked by the enemy, it is not his business to be firing himself; on the contrary, his constant occupation should be to see that his men do their duty well, and that they do not throw away

their

their fhot idly. If he perceive their ardour cooling in the midft of the attack, he fhould animate them by his voice ; and if he fees the enemy make a greater progrefs on one fide than on the other, let him weaken one to ftrengthen the other. I know this movement is fometimes dangerous, and that it would be better to have a fmall referve to make ufe of, as occafion may require : but can an officer, who has but a little detachment, fcarce fufficient to line the parapet two deep, can he take away twelve or fifteen men to make a referve ?

If the enemy fucceeds fo far as to make a breach and gain the parapet, two or three trees muft be immediately thrown into the breach, with all their branches*, and he muft receive the enemy with fixed bayonets.

One

* Provided the trees are to be found.—The true meaning is, that you are to ftop the breach with trees or ftones, or whatever you can, and make the entrance as difficult as poffible ; and on this account, it is not amifs to provide every field fort with fome materials for this purpofe.

One may alfo, as I mentioned in the
foregoing chapter, throw handfuls of
lime or afhes in their eyes, which will
foon oblige them to return to the ditch.
Laftly, if the true precaution has been
taken of furnifhing the hindermoft rank
with long arms†, and if eight or ten
men have been pofted in the ditch, fo
as to come round the redoubt, on right
and left, and take the enemy in flank,
one need not fear that they will eafily
mafter the place, or that their enterprize will coft them little.

But if it be the paffage of a river, or
a ford, that is to be defended, after
throwing, as I faid above, feveral trees
with all their branches on the bank, he
muft there wait refolutely for the enemy,
and keep up a fmart fire. If they attempt to come down the ftream in boats,
a good many grenades muft be thrown
among them ; one may alfo fire upon
them with large bird fhot, becaufe fuch
fhot fcattering a good deal, and wound-
ing fome in the eyes, fome in the face
or belly, troubled with fo many wounds,
which,

† Such as pikes.

which, though fmall, are neverthelefs
painful ; the foldiers who have received
them, will not fail to raife fuch confufion
in their party, as may make their
project mifcarry.

If, in the defence of a poft, which
has fome natural ftrength, and which
has been fortified according to the rules
I have given, the party fhould be forced
in the firft intrenchment, the foldiers
fhould retire into the ground floor, and
range themfelves at the loop holes ; in
that inftant two men fhould feize, as I
faid, on the two jaums of the door, to
ftop the enemy with their bayonets.

But if it fhould happen, that the foldiers, placed in the ground floor, fhould
be driven from thence alfo, one ought
not think that the enemy was therefore
become mafters of the poft ; thefe men,
forced below, fhould go up to the next
ftory with ladders, fuppofing the ftairs
were broken down ; they muft draw up
the ladders after them, and place themfelves at the holes, which fhould have
been made in the floor. If this ftory
fhould be low enough to reach the ene-

T   my

my through it with the bayonets, a single man will be sufficient to each opening; otherwise there must be two, who must not fire till they can almost touch the enemy with their pieces; these must also be ordered to pour down great tubs full of water, which must have been previously provided there, in order to spill down through the holes in the floor on the enemy that are masters below. This trick, though it may appear singular, is one of the most disagreeable that can be opposed to the assailants; for at the same time that it wets their arms, powder, and their cloaths, it hinders them to see what is passing over their heads, and frustrates any attempts they might make to set the house on fire. If, however, they should penetrate into a room, they must not be suffered to form, or reinforce themselves there, but they must be fallen upon sword in hand, or with fixed bayonets, and, by dint of bad usage, make them renounce the attack. The example I am going to relate proves, that the enemy is always

obliged

obliged to come to this pass when they
have to do with brave men.

During the war in Italy in 1705, M.
the Chevalier Follard being to defend a
little house or cottage, called Bouline,
near Brescia, where he commanded four
companies of grenadiers : this officer
was attacked by all the chosen troops of
Prince Eugene's army, who, after firing several cannon shot, and penetrating into the court of the house, were
forced to retire. " M. le Prince de Wirtemberg," says this author *, " who
" feared that we should be succoured,
" thought,. that in making himself
" master of a dove-house, from whence
" there was a hot fire kept up, the rest
" of the cottage would not hold out,
" upon which he attacked it ; and as
" our soldiers had carried off the door
" to make fire of, the officer who de-
" fended the lower part being wounded,
" could not withstand the fire that they
" made into this door, so was made
" prisoner ; there were seven grena-
" diers in the upper part of the dove-
T 2   " house,

* Commentaries on Polybius, tom. 5.

" houfe, who were alfo fummoned to
" furrender; but thefe imagining them-
" felves too well pofted to be foon re-
" duced to that neceffity, they an-
" fwered boldly, that they would not
" give up till the pears were ripe, and
" ready to fall; and that they were very
" capable of proving it. In effect, they
" continued to gall the enemy with
" their mufquetry, and did not quit
" their dove-houfe till the Prince de
" Wirtemberg retired, leaving the place
" covered with his dead."
This defence, which does great honour to M. Follard, and to his brave
captains who feconded him, is a good
leffon for young officers. M. le Chevalier de Clairac * gives another example, which is no lefs inftructive. In
1742, this officer, marching in the high
Palatinate of Bavaria, with a certain
number of people, perceived that he
was purfued by a troop of Huffars and
Pandours, who might attack him with
advantage; having examined the different

* See his Treatife of light Fortifications,
Chap. III.

ferent avenues of the village of Vurz,
where he then was, he barricaded them
with waggons, having taken off one or
two wheels from each, and with trunks
of trees, ladders, &c. he also raised a
banquette along the walls of the church-
yard, where he made a ftand with his
baggage and his followers, confidering
the church as a citadel, having broke
loop-holes through the door, and the
fteeple as an enclofure for his laft retreat; but there were two houfes that
almoft touched the church-yard wall,
and as they were built on lower ground,
the tops of their walls were no higher
than that, which ferved him as a parapet; neverthelefs he did not chufe to
open thefe houfes; but as he was obliged to have a communication with
them, in order to avoid being fired
upon from thence, and alfo to ufe them
as flankers, he contrived to make com-
munications, like bridges, from the
wall of his intrenchments to the roofs
of the houfes; and having barricaded
the doors and windows of the ground
floors, he pofts guards in them; but
thefe

thefe precautions were ufelefs. The Huffars, tired with watching him, fell back towards their army ; and M. de Clairac retired to Tirs-chen-reit, whither he was going.

Thefe examples, which prove what great refources a well-informed genius will derive from courage, fhew alfo to what a pitch the defence of an intrenched houfe may be carried, under the conduct of a determined refolution. The only means whereby the enemy may eafily force it, is to batter it down with cannon. If they once take this method, I fee no poffibility of their holding out long, unlefs, after it is quite down, they can contrive to range themfelves about the intrenchments.

Peafants houfes are generally fo ill built, and every cannon-ball is likely to make fo great a breach, that the defenders muft expect, in the end, to be buried under the ruins. The only methods then to efcape perdition, is either to capitulate, or to fally out brifkly on the enemy when he leaft expects it. The firft method cannot be thought of,

only

only upon condition of obtaining the
honours of war; which are, to march
out by beat of drum to return to the
army, whither you are to defire to be
efcorted and conducted by the fhorteft
road *. If the enemy will grant no capitulation, as the condition of foldiers,
prifoners of war, is always more grievous
and unhappy than death itfelf, fo death
would be preferable to it, if one had
not ftill the refource left, which experience proves to be almoft certain, of
faving yourfelf by a bold fally. The
neceffity one is then under of conquering, transforms the brave man into a
defperado, and opens him a paffage either to his army, or to a neighbouring
poft. It was by a *manœuvre* of this nature, that Count Saxe (afterwards
Marfhal-General of the French armies)
efcaped from Crachnitz, a village in
Poland, where 800 of the enemy's
horfe

* To ftipulate to be conducted by the fhorteft
road is a very material article; a General, who
forgot this circumftance in his capitulation at
Oftend laft war, was marched half round Flanders
with his garrifon, inftead of being conducted by
the fhorteft road.

horſe ſet upon him, and 18 of his followers, with intention to take him.
This Prince reſiſted them a long while
in the chambers of an inn at this place,
when ſeeing himſelf unable to hold out
any longer, he ſallied out unexpectedly
in the night, ſword in hand, cut his
way through the midſt of the enemy's
guards, and retired to Sandomir, where
he had a Saxon garriſon.

When the reſolution is taken to abandon a poſt, which can be no longer
maintained, you ſhould continue a very
ſmart fire till the inſtant the ſally is
made; and, in the mean while, remove
with the leaſt noiſe poſſible, the barricade from the door by which you are
to iſſue; when that is done, aſſemble
ſpeedily all your party on the ground
floor, march out in the cloſeſt order
poſſible, and with fixed bayonets drive
rapidly towards that part you have perceived to be leaſt guarded. "One ſhould
" never wait for day-light," ſays the
Chevalier Follard *, " to make theſe
" ſallies, which can only be ſucceſsful
" in

* Comment. on Polyb. Tom. V.

" In a dark night, by the opportuni-
" ty it affords of concealing from the
" enemy the road by which the retreat
" is made." For this reason you muſt
clear your way with your ſwords, and
not ſuffer a ſhot to be fired, leſt the
enemy ſhould direct their ſtrength to
the place where they hear the noiſe.

M. the Baron de Travers gives us
alſo a good leſſon on this ſubject. " To
" avoid being met by the enemy," ſays
this author *, " one ſhould always take
" a road quite different from that
" which the enemy might ſuppoſe we
" did take, and which ſhould appear
" to be what we ought to take: a ſmall
" party can hide themſelves any where;
" and as it is not common to ſeek thoſe
" places near the enemy, thoſe there-
" fore are the ſafeſt; there they ſhould
" paſs the day, and take another road
" under cover of the night."

But if the poſt be conſiderable, ſuch
as a village or borough, whereof the
defence is committed to an officer, he
may kill a great many of the enemy

U   before

before he is obliged to make his retreat. When his smaller posts have
held out as long as possible, he will
make them fall back to his principal
one, still fighting from street to street,
and from trench to trench. But in
order that the solders may execute these
manœuvres easily, he must, as I have
before said, have exercised them therein
before-hand. In a defence of this nature, the commandant should also ob-
serve with great attention all the motions of the enemy, so as to distinguish
a feint from a real attack.

Though the enemy should succeed so
far as to force all the intrenchments,
and to get footing in the village itself,
it must not therefore be taken for
granted, that he has gained the victory.
An officer, retired to his principal post,
may begin a-new to give him such a
reception, as I have particularized in
the method of defending single houses,
so as to disgust him entirely with his
enterprise, and oblige him to retire
again.

I have

I have said, and shall repeat it again, the defence of a post, of a village, or even of a city, is so easy, that I cannot comprehend why they do not hold out longer than they commonly do. There is nothing necessary for it, but resolution, vigilance, to know how to make the most advantage of the ground, and to persuade the soldiers that nothing but downright cowardice can let the enemy penetrate into the place. The example of Cremona in 1702, will be an everlasting proof of what determined courage can do, and will teach posterity that, though the enemy should be possessed of half the ramparts, and of a part of the town itself, they are not yet entire masters of the place.

Prince Eugene having formed a design to surprize this town, which was our head-quarters, defended by a garrison of French and Irish; some thousand Austrians were introduced there by a priest. These troops immediately made themselves masters of two gates, and of a great part of the town: the garrison buried in sleep, started up in surprize,

and were obliged to fight in their fhirts ;
but the French officers directed their
manœuvres with fo much prudence, that
they repulfed the Imperialifts, from place
to place, from ftreet to ftreet, and obliged
Prince Eugene to abandon the part of
the town, and the ramparts that he was
in poffeffion of.

What then hinders us, now-a-days,
from defending villages where we are
pofted, in like manner, and from dif-
puting the ground inch by inch ; efpecially when a church, or *chateau*, is fecured as a fure retreat, fit to make a
good defence, and to obtain an honourable capitulation ? This is eafy, and yet
we fee few or no examples ; becaufe we
do not apply ourfelves fufficiently to
learn the caufes of the difafters, which
our predeceffors fell into, for want of
knowing better.

One may judge, from what I have
faid on the defence of pofts, that there
is nothing more eafy than to maintain
them ; thofe who attack, having nothing fupernatural in them, but are the
fame

fame fort of men with thofe who are attacked.

A determined commander, who is jealous of his reputation, and who has learned, by ftudy, to make ufe of his talents, dares, like Leonidas, with 300 men, defend the ftreight of Thermopile, againft a whole army; and as a modern philofopher fays, had rather perifh glorioufly, than be guilty of a cowardly action. In fact, an able commander is never aftonifhed at the numbers of the enemy; in a houfe, a village, or even in a town, he may oppofe devices to them, which will always fupply his defect of forces.

I have feen, during the laft war in Piedmont and Italy, intrenchments and pofts that have withftood the firft and the ftrongeft attacks of the affailants; and which have been given up or abandoned in fome following attacks, tho' not near fo hot as the former: from whence comes this? It is becaufe the officer that is placed in the poft, dare not abandon it on the firft attack; but he defends himfelf and repulfes the
enemy;

enemy; becaufe if he fuffered himfelf
to be forced, he and his men would be
all put to the fword. But fhould the
enemy return, a commandant imagines
that he has nothing to reproach himfelf
with, becaufe he defended it fome time;
and then he either retires, or furrenders.

But fince he was able to repulfe the
enemy immediately when they came
frefh, and in good order, to attack him;
with how much more facility may he repeat the fame, when they return har-
raffed and fatigued, and in a condition
much lefs to be dreaded than the firft
time? Does not the caufe of this come
from, not exciting fufficiently the emulation of military perfons? An officer
who is not countenanced, and who is
never affured of the leaft reward, neg-
lects himfelf, and thinks lefs of acquiring glory, which, though obtained by
a brave action, is for the moft part un-
known, than of enjoying quietly a common reputation.

"We commend greatly," fays Mon-
"fieur Follard *, "and I believe we
"can,

* Commentaries on Polybius, Tom. 5.

" cannot too much either commend or
" reward those who make a noble de-
" fence in the posts that are committed
" to their charge. The reason of this
" is, that the recompences for these
" kind of actions being much greater
" than those given for others, excite
" and animate officers to defend a post
" vigorously to the last extremity.—
" But if the requital should be pro-
" portioned to the action, then he that
" has done nothing worthy of a brave
" man, but has surrendered in a cow-
" ardly manner, ought to be stripped
" of his arms, and put to death with-
" out mercy. This was a law with
" the Romans * ; but the General, on
" his part, should be attentive to leave
" the officer no room to complain, and
" that he should be furnished with eve-
" ry thing requisite for his defence.
" It is not necessary, says the same au-
" thor in another place, that an officer
" who is fixed in a post should be over-
" ready

* And in France also ; by an ordonnance of
the 20th of July, 1714, it is forbiden, under pain
of death, to quit or desert a post,

" ready to come to action; but that he
" fhall always refift when he is preffed,
" and that he fhould die, rather than
" abandon his intrenchment."

Ancient and modern hiftory furnifh
but few examples of pofts being well
defended ; and it feems as if the military authors had agreed not to fpeak
of actions of this nature. Neverthelefs, it is not to be doubted, but that
in the different wars which France has
waged, chiefly under the reign of Henry the Fourth, and of his Succeffor,
in times when the armies were not near
fo numerous as they are now-a-days,
there were officers whofe actions de-
ferved to be remembered in hiftory.
However, we don't find any body has
recorded them ; although the leffons
that might have been taken from them,
would have been as inftructive and as
agreeable to read, as thofe that have
been left us of the beft fortified places
of a ftate   If I am furprifed to fee in
1604 a hundred thoufand men perifh
before Oftend, and their General *,
                                    with

* Archduke Albert.

With the remains of his army, not able to mafter the place, till after a fiege of three years; I am no lefs afto-nifhed to fee Charles XII. King of Sweden, in 1713, with feven or eight officers, and fome fervants, defend a wooden houfe near Bendar againft twen-ty thoufand Turks or Tartars. We find in many hiftorians the defence of this houfe, becaufe it was made by a crowned head; but great actions, let the authors be who they will, ought not to be buried in oblivion. They not only pique the emulation of offi-cers, who always find matter of in-ftruction in them; but they are ho-nourable to thofe who perform them, to the corps they belong to, and even to their nation. I am forry I could not collect a greater number of au-thentick ones, for it would have been a great pleafure to me, to have em-bellifhed this little work with them.

X   CHAP.

CHAP. X.

Of the attack of posts.

THOUGH to take a post from an
enemy may be always a difficult
task, if those who are to defend it know
their business; however, the way to
succeed is, either by a rapid and sudden
attack, or by stratagem.

One ought never to form the project
of an attack upon simple speculation;
because our imagination often makes
us think things feasible, which, when
we come to the proof, we find impossible to execute. When an action of
this kind is proposed to be undertaken,
one should form a just idea of it; examine
separately every part of it, and the various means that are to be made use of,
and compare them together, to see if
they correspond with each other, and
answer to the general end: lastly, one
should take such just measures, that, if
I may be allowed to say so, you may
be insured of success before you begin.

As

As it is not cuſtomary in an army to
chuſe a particular officer of foot to attack an intrenched poſt, unleſs he offers his ſervice; an officer ought not to
embark in ſuch an enterpriſe, without
examining into the means of ſucceed-
ing, and being able to ſhew a plan of
his project to the General, in order, if
he approves of it, to gain his conſent
to put it in execution. In caſe the General likes the plan, the officer is to de-
ſire leave to take a nearer view of the
poſt, with one or two men, in order to
take more exact meaſures for the execution. I ſay he ought to aſk leave to
go and view the poſt, to the end that
if he ſhould be diſcovered and taken
priſoner, he may be owned, and reclaimed.

How to reconnoitre a poſt.

An officer who goes to take a near
view of the poſt he intends to attack,
ſhould go out at the beginning of a
dark night, and give thoſe, that go
with him as aſſiſtants, inſtructions how

to act in every circumftance; fuch as,
to examine well every place through
which they pafs; to approach the poft,
by fearching with long poles, left there
fhould be any traps or ditches covered
over, into which they might fall; and
to ftick large branches of trees, with
the leaves on, at thofe they difcover, fo
as to guide them when they return to
the attack; to take particular notice of
the pofition of the centries, their di-
ftances from each other, and their number; to advance to the edge of the
ditch, to try the depth of the water with
their poles, or with a lead and line; to
fee whether the poft is *fraifed* * or palifaded, built of earth or fafcines, or
mafonry: in the laft cafe they are to
guefs, as near as they can, at its height,
to be able to proportion the length of
their fcaling-ladders. Laftly, to know
how many men the garrifon confifts of,
and in what they are negligent; if they
are

* A *fraife* is a pallifade laid horizontally, or
nearly fo, being half buried in the earth of the parapet; the other end fharpened, and pointing out
towards the enemy Palliffade, properly fo called,
is commonly fet perpendicular.

are likely to receive any fuccours, or if they have any cannon, &c.

It is upon the knowledge of all thefe circumftances, which one can examine into himfelf, or may learn from the report of deferters or peafants, that an officer may form the project of an attack.

If a perfon receives his inftructions only from the reports of others, he muft be cautious how he gives credit lightly to thofe, whom perhaps either a defire of betraying him, or the hopes of recompence, might have induced to throw themfelves in his way; on the contrary, he fhould queftion them feparately, write down what they fay, compare their depofitions, and judge afterwards what part of their intelligence may be true, or what falfe. Having taken thefe inftructions, the officer fhould return to his General, to inform him of his difcoveries, and receive his laft orders for the attack; for the foldiers that are to fecond him, and thofe who are to march to fupport him.

Of

Of the choice of soldiers.

The choice of men who are to march to the attack of a post is a thing so essential, that the success of the enterprize depends upon it. Therefore none should be chosen but willing and bold soldiers, and who are not rash or heedless, and none who have colds; for a man, who without attention to the orders of his officer, will suffer himself to be led by the heat of his zeal, or who by coughing or spitting, discovers the march of his party to the enemy's centries, may make the best concerted project fail. As to those that are to support him, he is to take them according to their turn for guard, or for detachment, as the General shall think proper.

Of Dispositions.

The dispositions for an attack ought to be relative to the discoveries that have been made; so that one should not be obliged to return in the middle of the execution.

The

The men being chosen, they are to
be inspected, to see that they want nothing that may contribute to their success; I say, that may contribute to their
success : because, if the post is fortified
with an intrenchment of earth, or of
fascines, the two first ranks should be provided, besides their arms, with shovels
and pickaxes; if it is *fraised* and palli-
sadoed, they should have good hatchets;
and if it is faced with stone or brick-
work, they should carry scaling ladders.
All the soldiers also ought to be in their
waistcoats, so as to be more at liberty;
and they should have paper cockades,
that they may know one another in the
dark : after this inspection, the following disposition is to be made.

If the intention is to make one or
two true attacks, and as many false
ones, the chosen men are to be formed
into as many platoons as there are to be
true attacks; and the others appointed to
support them, are to make the false ones,
in order to divide the enemy and their
fire. Then a man, capable of commanding, should head each platoon; and ob-
serve

ferve that thefe officers, as far as poffible, may be the fame who were at the examining the fituation with him ; fo that each of them may be able to guide his divifion.

Thefe officers are to be ordered to march together, till they come to the place appointed for their feparation ; whence they are to go each to their feveral ftations near the poft, where they are to lie down on their bellies, and wait for the fignal of attack to leap into the ditch, and fcale the poft.

Of Guides.

If an officer is to be conducted to a poft, by guides or fpies, he fhould firft queftion them carefully, fo as to draw from them as much ufeful information as he can, particularly touching the nature of the road, by which they intend to conduct him. The reafon of which is, becaufe one fometimes meets with filly fellows, who, animated with the love of lucre or otherwife, believe that they can conduct a body of men eafily,

eafily, at the fame time that they are totally incapable, and truft folely to their good will. But if he finds them fufficiently qualified for the purpofe, he fhould ufe all poffible means to be affured of their fidelity, by making them dread the total deftruction of their houfes, and pillage of all their goods, if they lead his troops into any traps. He may alfo demand their wives or children, as pledges for their good behaviour; and at the time of marching, they fhould be placed between the corporals of the firft rank, fecured with a fmall chain or cord. This laft precaution is the more neceffary, as traitors have been often found, who under pretence of helping to furprize a poft, have conducted a body of men into a cutthroat place in the middle of the night, and flipped off themfelves in the midft of the fray. On one hand, a recompence proportioned to the fervice fhould be offered to thofe people, in cafe their conduct fhould be good; on the other hand, they fhould be threatened with the moft fevere punifhment if it fhould be bad.

Y

Of the March.

Night being the beſt time to ſteal upon a poſt, care muſt be taken to ſet out time enough to get near the place an hour or two before day, provided that it is not moon light at the ſame time ; for if it be, the attack ſhould be deferred, if poſſible, till the moon gets under a cloud, and then ſeize the moment of obſcurity to begin the work : the ſoldiers ſhould march, two by two, as lightly and ſilently as poſſible, eſpecially when they are to paſs between two, of the enemy's centries, they are to be forbid to talk, cough, ſpit, or ſmoak.

When the detachment is arrived at the place where the platoons are to ſeparate, the officers of theſe diviſions are to repair, with them, to the places I before-mentioned, for them to lie down, to wait the ſignal ; obſerving, that the places where they lie in ambuſh, ſhould be oppoſite the ſalient angles of the intrenchments, ſeeing that theſe are the

ſpaces

ſpaces leaſt defended by the enemy's muſketry.

If whilſt the parties are on their march, or lying in ambuſh, they ſhould happen to meet the enemy's patrole, they are not to be alarmed on that account, or make the leaſt motion, becauſe it might ſpoil the whole enterprize: they ſhould only remain hidden, and totally ſilent; ſo that the patrole may paſs by without ſeeing them, and that they may afterwards purſue their own deſign.

The attack of a common redoubt.

If the poſt to be attacked is a redoubt, with a dry ditch and a parapet of earth; the two firſt ranks of each diviſion, as I ſaid before, are to be provided with ſhovels and pick-axes, and are to ſling their muſkets. Things being thus prepared, as ſoon as the chiefs ſee, or hear the ſignal, all the diviſions are to riſe, and march with ſpeed, to leap into the ditch at the ſame inſtant; I ſay at the ſame inſtant, becauſe it ſhould be a maxim in the attack of a poſt, for all

to

to fall on at once. When the firſt rank
have leaped into the ditch, the ſecond
ſhould ſtop a moment, leſt they ſhould
leap on the backs of the firſt, and throw
themſelves on their bayonets. Theſe
two firſt ranks being got into the ditch,
they are immediately to undermine the
angles of the *ſcarpe* or ſlope, and the
parapet of the redoubt, to facilitate the
climbing up of the reſt of the party.
The officers of each diviſion are to take
care, that the ſoldiers armed with their
muſkets, who have alſo got into the
ditch, may not obſtruct the workmen,
but that they protect them, by preſent-
ing to the right and left; and that they
are always ready to repulſe the enemy
that may have been poſted in the ditch.
If the parapet be *fraiſed*, they are to
cut away with their hatchets, as many
of theſe pointed poſts as will leave a
ſufficient paſſage; and when the breach
is made, the workmen are to lay down
their tools, handle their arms, and
mount all at once with fixed bayonets,
and fall upon the enemy, crying out,
kill, kill.

When

When a body of men march againſt
a redoubt, or any other poſt, with in-
tent to ſurprize it, the commanding officer ought always to make his attack
on that ſide, which may have communication with ſome other more confiderable poſts, in order to cut off this
communication ; for people who ſee
themſelves warmly attacked, and have
no hopes either of retreat or ſuccour, will
very ſoon aſk for quarter.

The attack of intrenchments with a revète-
ment, i. e. *faced with maſonry.*

Though the attack of poſts, whoſe
ſcarpe or ſlope,* and parapet are faced
with brick or ſtone-work, can only be
made by *eſcalade* ; neverthelefs it ſuc-
ceeds, if it is briſkly ſurrounded, and
well ſupported.

An officer who intends to attack a
poſt of this kind, muſt take care that
the

* The *ſcarpe* or ſlope, reaches from the bottom
of the ditch, up to the ordinary level of the
ground ; and the *parapet* is the ſuper-addition
lying above the *berme,* which marks the ſurface
or level of the ground.

the ladders he intends to ufe, fhall be
rather too long than too fhort ; and let
only the ftrongeft foldiers carry them.
Thefe foldiers are to carry them with
their left arms thruft through the fecond
ftep ; they muft keep them upright,
and clofe to their fides ; and hold them
fo fhort, as not to come near the ground,
to avoid diflocating their fhoulders,
when they leap into the ditch. The
firft ranks of each divifion, being thus
provided with ladders, are to fet out at
the firft fignal, with the reft of the
party, and are to march boldly with
their pieces flung, and their fwords in
their right hands, and leap into the
ditch : when they are down, they are
to fix their ladders againft the wall, rather towards the falient angles than
againft the curtins ; becaufe there the
enemy are by much the weakeft ; and
care muft be taken to fix the ladders
but a foot from each other, and not to
give them too much or too little flope ;
as in one cafe they may be eafily over-
turned, and in the other they will be
too weak to bear the men.

When

When the ladders are fixed, thofe
that carried them are to mount directly,
and to be followed immediately by all
the reft, to fall on the enemy fword in
hand. If he that gets up firft fhould
be knocked down, the next man muft
take care not to let himfelf be beaten
down by the falling body ; to avoid
which, he muft endeavour to make him
pafs on one fide, between the two ladders, and then climb up as quick as
he can, not to give the enemy time to
reload.

As the foldiers who mount firft are
likely to be knocked down by the
enemy's fire, and as their falling may
happen to make the attack fail, I think
it would be right to give them a light
cuirafs or breaft-plate ; becaufe, as foon
as thefe have got in, the reft will eafily
follow. Some people may think this an
ufelefs precaution ; but is it better then
to have all your people expofed to be
knocked on the head in the ditch, than
to carry the poft with more certainty
and lefs danger ?

The

The fuccefs of an attack by *efcalade*
is infallible, if they mount with fpeed
on the four fides ; if they take care to
fhower in grenades in abundance ; and
if they are fupported by fome companies of grenadiers, and by fome pickets,
who will draw off and divide the enemy's
fire and attention.

Of the paffage of a ditch full of water.

If the ditch of the poft to be attacked
is full of water, and only takes a man
up to his belly, that need not hinder
their jumping into it, and carrying on
the attack, as defcribed in the foregoing
fection ; but if it is too deep to be
paffed by wading, the foldiers of each
platoon muft carry fafcines or faggots,
of flender branches, made as thick as
poffible, and tied very tight, to fill up
the ditch, or render it fo far fordable,
that the affailants may get to the parapet, either to undermine it, or to
fcale it.

Some authors recommend, for this
purpofe, cafks filled with earth ; and
M. de

M. de Follard, facks filled with dung *,
or litter, of five feet diameter; but I have
found, by many trials, that the cafks are
very difficult to roll, efpecially if the
ground is uneven, and you have a confiderable way to move them; that it is difficult to fill up the ditch with them, becaufe their folidity makes the water rife
higher and higher: that facks of earth,
or of dung, cannot be rolled, on account of their weight; that they burft
in the carriage, fpill their contents,
make the ford very muddy, raife it but
little, and leave it ftill difficult to be
paffed. Therefore fafcines are preferable to all thefe, becaufe the foldiers
can carry them before them, where they
ferve to cover them from mufket-fhot;
and being light, they do not retard
their march. All thefe fafcines, which
may be handed from one man to another, and thrown into the water, will
foon fill up the ditch, fo as to make
a paffable ford.

M. Follard gives us another method
of paffing a wet ditch; which is, to

Z   make

make frames of feven or eight feet broad,
by ten or twelve long. " Thefe frames,"
fays he *, " confift of three bars of
" wood, with crofs-bars, in the manner
" of a hurdle, and well mortoifed;
" planks fhould be nailed on the top,
" and a grapple faftened to one end,
" to cling to the fafcines of the enemy's
" intrenchment."

But M. Follard has not told us how
thefe bridges are to be carried to the
ditch, or how an officer is to get them
made.

*Ways to counter-act the other contri-
vances.*

If the approaches of a poft are defended by *chevaux de frife*, the firft and
fecond ranks of each platoon fhould cut
away the fpikes with hatchets, or they
may hale them forward, and throw
them afide, with an iron grapple faftened
at the end of a rope; if by an *abbatis*,
they fhould throw fafcines or great
faggots on the points, and over the
branches, by which means the foldiers
will

* Commentaries on Polybius, tom. V.

will be able to pass easily over it: also,
if these *abbatis* are double or triple,
they may be set on fire, by throwing
well dried faggots, lighted at one end,
into the middle of them. If this last
proposition is to be executed, as soon
as the lighted faggots are thrown on
the *abbatis*, the soldiers should retire to
a certain distance, so that the enemy
may not have an opportunity to level
their shot at them by the light of the
fire ; and they also should be so posted,
as to be able to fire at those of the
enemy who shall endeavour to extin-
guish the flames.

But, lastly, if the avenues to a post
are defended with caltrops, they must
be swept away, by dragging one or
two trees, with all their leaves on, over
the ground where they have been laid.

The attack of a chateau, or of a house.

The approaches of a *chateau*, or a
house, are to be attacked in the same
manner as those of a detached or single
post: after these are taken, trial is to

Z 2   be

be made to gain the upper part, by efcalade, and to deftroy the befieged by
knocking them down with tiles: but
if the enemy have uncovered the poft,
in order to prevent this; then hand grenades fhould be thrown in at the door
and windows as faft as poffible; alfo a
quantity of dry fafcines, with lighted
faggots dipped in rofin, fhould be thrown
in, to endeavour to ftifle or burn out
the enemy. If it fhould be windy weather, advantage may be taken thereof,
to blow the flames towards the houfe;
and attempts fhould be made to ftop
the loop holes the enemy may have
made in the walls with fand bags, in
order to fet about *fapping,* or undermining the angles of the building; but
if the affailants have any cannon,
the work may be fhortened, by pointing them at the angles of the poft.
Inftead of cannon a great beam may
be fufpended with ropes from a triangle,
made with three ftrong poles, in imi-
tation of the *battering ram* of the Romans. This beam forcibly pufhed againft
the walls, will foon make a large breach;
but

but care muft be taken only to mount it in a dark night, fo that the enemy may not obftruct the work, dy killing the men that are employed about it. If it is a glorious part to come off with honour from an attack of this kind, it is not lefs fo to execute it with the lofs of only a few men. One cannot be too fav-ing of the blood of foldiers; and a know-ing officer ought never to neglect the means that can contribute to preferve his men. The comparifon of two examples that I will relate, will fhew the impor-tance of what I advance. During the two fieges of Barcelona made by M. de Vendome in 1697, and M. de Barwick in 1713, the firft of thofe generals at-tacked fword in hand the convent of capuchins fituated without the town, with feveral detachments of foot, and made himfelf mafter of it in three hours time, with the lofs of feventeen hun-dred men. Marfhal Barwick attacked the fame convent in 1713: the enemy were equally intrenched, and reckoned to make the French pay as dear for the victory as they had done before; but

this

this general opening a kind of trench
before the convent, the enemy, who
did not expect to be attacked in form,
furrendered at difcretion, after twenty-
four hours refiftance. I fubmit it to the
judgment of my military readers, to
decide which of thefe examples is beft
to follow.

The attack of a village.

The preparations for the attack of
a village, or any other extenfive poft,
are the fame with thofe mentioned in
the former part of this chapter for
thofe pofts that ftand alone. But as
thefe kinds of attacks are always more
difficult than others, on account of the
various devices that may be oppofed
to every attempt; an officer fhould not
begin his movements,* till he knows
the ftrength of the intrenchments, the
fituation of the little pofts, the obftacles

* That is, he fhould not move until he has
got the beft information the nature of the thing
admits of; an officer muft be cautious in conftructing this fentence, not to draw hence reafons
for his inactivity, under pretence of his not having fufficient intelligence.

ftacles he may meet with in every
ftreet, and how the inhabitants ftand
affected towards the garrifon.

If an officer takes this information
from the people of the country, he
fhould affect a great indifference in his
enquiries, that they may not fufpect his
defign, and communicate it to the enemy, who by that means may take precautions to overturn his project.   He
fhould alfo endeavour to be well affured of their truth, as I faid before,
by comparing the reports of peafants,
deferters, and what he knows, or has
feen himfelf, all together, in order to
find out the moft probable.   When he
knows the enemy's fituation, he is then
to make his difpofitions for the attacks, and muft point out the duty to
the officers of each platoon, as well to
thofe of the falfe attacks, as to thofe of
the true.

The true attacks are to be made at
the places that are moft difficult of accefs, becaufe here the enemy, confiding
in the ftrength of the fituation, are the
least

leaft on their guard *. They may alfo
be made upon the houfes fituated at
the entrance of the ftreets, becaufe,
when you are in poffeffion, it is an eafy
matter to break through the walls of
one houfe into another; and being poffeffed of the houfes, it will be eafy
to drive the enemy out of the ftreets
even with ftones, and oblige them to
take to their laft intrenchment.

If the war is in an enemy's country,
which you do not chufe to fpare, it
would be an eafy matter to fet fire to the
four corners of the village, and oblige
the befieged to furrender themfelves
prefently; but befides the inhumanity
of ufing means that tend to lay wafte a
whole country, it is likewife very dangerous to throw all the inhabitants of
the open country into defpair, becaufe
then flying into the woods, they form into
bodies, and fpread about every where,

to

* People always guard againft probabilities;
therefore, as Cardinal de Retz fays, probabilities
feldom come to pafs:—at the late conquefts of
Cape Breton and Quebec, the fuccefsful landings
were made at places, deemed by the enemy (the
French) to be inacceffible.

to knock the ftraggling foldiers on the
head, murder the futtlers, hinder the
peafants to carry any provifion to the
camp, and ravage the whole country.
M. Follard * fpeaking on this fubject,
fays, " we faw during the war in 1688,
" fifteen hundred Barbetts of the valley
" of St. Martin, keep forty batallions
" of our troops in awe, through the
" whole extent of the valley of Pra-
" gelas, where the Bifan runs in the bot-
" tom between two very high moun-
" tains, of moft difficult accefs, and
" where each party guarded their own
" fide " . . . . Thefe mountaineers came
down fometimes when they faw our con-
voys in motion, and attacked them.
At the fame time they had fcarcely ten
or twelve men where we had entire
corps.

It appears from the example I have
now mentioned, that it is very impru-
dent to mafter confiderable pofts, fuch
as villages, by fetting them on fire; and
that it is much better to take them by
fmart attacks.

A a   An

* Commentaries on Polybius, tom. IV

An officer who commands an expedition of this kind, fhould not attach himfelf obftinately to one fingle attack; for the falfe ones may become true, and he ought to know the fuccefs of each, fo as not to throw away men to open a paffage on one fide, whilft perhaps it may be already open on the other.

When the affailants have penetrated the village, the commanders of each divifion fhould be attentive to leave fmall detachments at every church, and at every ftrong and tenable place fit for the bulk of their party, in cafe they fhould be repulfed,

They fhould be very watchful that the foldiers do not fcatter about to pillage the inhabitants houfes. Detachments have been often driven back out of a town or village, by neglecting this precaution.

Three days after the furprize of Cremona, in 1702, fome German foldiers were ftill found in the cellars, where they had got drunk, and were greatly aftonifhed, when they were told that they muft quit thofe lovely retreats.

An

An officer, who would avoid so dangerous a disorder, should make it death
for a soldier to stir from his party, and
should post serjants in the rear of each
division, to prevent any man from stay-
ing behind.

If they find any cavalry drawn up in
the markets, or open places of the village, the besiegers are to stop, and stand
firm at the heads of the streets opening
into the place or market: some of them
are to get into the houses next the corners, to fire on the enemy through the
windows; and if they find it disorders
them, they should immediately charge
them with fixed bayonets, to oblige
them to surrender.

Lastly, if the internal parts of the
village be defended by cannon, the
troops should march with speed to pos-
sess themselves of them, to spike them
up, or to turn them on the enemy, or
against their chief post in the village.

One may judge, from all that I have
said on the surprizing and seizing of
posts, that those actions, tho' difficult,
are not impossible, when the means intended

A a 2

tended to be employed in the execution
are judiciously combined, and well confidered. Thefe ways are eafy to be
imagined ; neverthelefs few examples
of fuch actions appear, becaufe people
do not apply themfelves enough to this
part of war ; wherein to fucceed, it requires good fenfe, great courage, a head
for ftratagem, a daring fpirit, a ready
execution, and a cautious forefight.

We find in antiquity, an example of
an attack, which by the recital of its
circumftances, may be of great fervice
to regimental officers : I take it from
the feventh book of Polybius. "The
"blockade of Sardis by Antiochus the
"Great," fays the tranflator * of that
author, "held for two years, when La-
"goras of Crete, a man fkilled in the
"art of war, put an end to it in this
"manner" He had remarked that the
ftrongeft places were often taken with
the moft eafe, owing to the negligence
of their defenders, who confiding in the
ftrength of the natural or artificial fortifications of their town, take very little

* Don Vincent Thuillier.

pains

pains to guard it. He knew alſo that places were ſometimes taken by an attack on their ſtrongeſt ſide, and where the defenders did not expect that the enemy would undertake any thing.

And though he ſaw plainly that Sardis was looked upon as an impregnable fortreſs, not to be attempted by aſſault, and that nothing but famine could oblige them to open their gates ; nevertheleſs, upon the foregoing conſiderations, he hoped to ſucceed. Difficulties only increaſed his application to think of all poſſible means of entering the place ; and obſerving that the part of the wall which joined the citadel to the town was not guarded, he formed a deſign to ſurprize it in this place ; the proof of this ſide's not being guarded, was thus : this wall was built on a very high and ſteep rock, at the bottom of which was a kind of deep pit, wherein the town's people were uſed to throw the dead carcaſſes of horſes and beaſts of burthen ; there vultures and all ſorts of carniverous birds aſſembled every day, and after eating their fill, never

failed

failed to reft themfelves either on the
rock or on the wall.

This was enough for our Cretan engineer, to fee that this place was for the
moft part of the time neglected, and
without any guard. Upon this fuppo-
fition, at night he went to the place,
and examined how he could approach
it, and where he might apply his ladders. Having found a fit place againft
one of the rocks, he imparted his defign and his difcovery to the king; who,
delighted therewith, exhorted Lagoras
to go on with his enterprize, and gave
him two other officers that he required,
and who feemed to have all the abilities and valour that his project de-
manded. Thefe three having confulted
together, only waited for a night, at
the end of which there fhould be no
moon light. When it came, they chofe
fifteen of the ftrongeft and ftouteft foldiers in the army to carry the ladders to
fcale the walls, and to run the fame ha-
zards with themfelves. They alfo chofe
thirty others to lie in ambufh in the
ditch, to affift thofe who fhould fcale
the

the wall, to break open a gate to which
they were to march. The king was to
order two thousand men to follow these,
and to favour the enterprize by marching the rest of his army to the opposite
side of the town. Every thing being
ready for the execution, as soon as the
moon was down, Lagoras and his people approached softly with their ladders,
and having scaled the rock, they came
to the gate which was near it, and broke
it open, having forced all whom they
met in their way; the gate being open,
the two thousand men entered, when
some cutting down all that opposed
them, while others set fire to the houses,
the town was pillaged and ruined in an
instant.

I intreat young officers, who shall read
this example, to consider well on this
attack. The diligence and attention of
Lagoras, who went himself to view the
proper place to fix his ladders, his dis-
cernment in his choice of the officers
and soldiers that were to aid him, and
the exact concord between the several
means that he used, are so many lessons
for

for thofe who fhall be tempted to undertake fuch an attack

'This of M de Roche-Fermoy, captain in the regiment of Bourbonnois,
performed before Charleroy, is no lefs
inftructive.

During the fiege of that place, at the
end of July, 1746, M. de Lautrec, then
lieutenant-general of the trenches, per-
ceiving that the taking the redoubt of
Marcinelle, which defends the lower
town, was abfolutely neceffary to
ftreighten the garrifon ; he ordered M.
de Roche-Fermoy, a brave and determined officer, to take a clofe view of
this important poft. This officer ac-
cepted the commiffion, fet out with a
fingle man at the beginning of the night,
paffed between the enemy's centries,
and advanced even upon the glacis of
the redoubt. Having founded the
water, which feemed very deep, he
found in one place five or fix feet,
and in another only four. He alfo faw
the poft was *fraifed* and palifaded, and
defended by feveral pieces of cannon,
and fifty Auftrians commanded by three
officers.

officers. All these obstacles were not sufficient to dishearten him ; he tied his cockade to some reeds opposite the place where the water was but four feet in the ditch, and in his way back left his coat at some distance from that, to be a guide to him on his return to attack the post. When he came back to the trenches, he reported his discoveries to M de Lautrec, who gave him forty chosen men, and ordered him to be supported by Monsieur de la Merliere, captain of grenadiers, who having marched to the opposite side of the redoubt, and drawing the fire of the garrison that way, favoured this enterprize. As soon as M. de Roche-Fermoy got near the glacis, he made all his men lie down close, to wait the signal, which was made an hour before day. Then this officer leaped into the ditch, ordering his men to put their cartteridge boxes on their heads, to preserve them from wet ; made them cut away with their hatchets as many of the *fraises* as were necessary to open a passage, climbed up the redoubt, and fell on the ene-

my

my with fixed bayonets, who being
furprized at fo fudden a vifit, fought
their fafety by flight; but M. de Roche-
Fermoy having ordered the draw-
bridge, that communicated with the
town, to be hoifted up, they were
obliged to furrender at difcretion. The
fmartnefs of this attack, the orders of
M. de Lautrec to make it fucceed, and
the activity with which the prince of
Conty directed the works of the fiege,
making the enemy fear their being
taken by affault, the Generals Beauford
and Halkert, who commanded, gave
up the place to the prince, and were
made prifoners. The next day M. de
Roche-Fermoy was prefented to his
highnefs, with all the Germans he had
taken; the prince greatly commended
his valour, and fent fo favourable an
account of his conduct to court, that
the king immediately fettled a penfion
upon him, to be paid out of the royal
treafury.

I confine myfelf to the examples that
I have given, on the fubject of *attacking pofts*, having found no more among
feveral

feveral that I have looked over, that were fufficiently authentic, or well enough explained to make ufe of; intelligent officers will fupply this defeſt by their own reflection. Thofe who really love their profeffion, will both gain knowledge, and make new difcoveries by application only. Certainly I fhould be flattered, if what I have given fhould be approved ; but much more fo, if the rules that I have propofed can contribute to the fuccefs of officers, fo as to diftinguifh them, that they may arrive at thofe high ranks which are the recompences of military virtues.

Of feizing pofts by ftratagem.

That part of military fcience which comprehends the furprizing of pofts, is little capable of being treated in a methodical manner. The particular intelligence of each officer, and feveral opportunities that happen by chance, are what commonly occafion the execution of thefe kind of actions.

　　　　　War

War being an art depending much
on *finesse* and ftratagem, there are a
number of precautions that efcape the
forefight of men in action, which a
fkilful enemy may obferve, and which
furnifh him with opportunities to make
good ftrokes Hiftory contains many
examples of the like actions, which are
only rare now-a days, becaufe we do
not fufficiently ftudy this part, wherein
is required an elevated genius, and a
combination of means relative to the
true pofition of the enemy, which one
ought always to view one's felf.

But alas! how many people lofe their
underftanding when they are obliged to
take a clofe view?

I diftinguifh two forts of furprizes in
war, the general, and the particular.
The firft are fuch as are undertaken
againft a camp, an army, or a fortified
town. As it is neceffary, in order to
fucceed in thefe, to take the precautions that have been already laid down
by able heads, and to be affifted by a
much great number of men than what
private officers commonly command; I
shall

shall not speak of them. I shall confine myself to those that may be performed by a small body of men, or for the success of which very extensive measures will not be required.

In all times stratagems were used in war, " says the author of the military
" dictionary, a work useful for all of-
" ficers. Frontinus under Trajan, and
" Polyenus under Antoninus, have
" wrote on the stratagems of great cap-
" tains, and even of illustrious women.
" . . . . Each general has his own, adds
" this author; there are some which
" owe their rise to time and place only,
" and which ought never to be neg-
" lected. Many people pretend that
" every thing is fair in war, and that
" you may procure the success of what
" you undertake, by any means what-
" soever. But authors, who have wrote
" on the *jus gentium*, or the rights of
" nations, do not agree on this head."

I will remark on this subject, that M. le Chevalier Follard thinks all stratagems equally good; tho' in the number of those that he relates, there are, according

according to him, fome wherein fince-
rity and greatnefs of foul fhine with
great luftre ; and others where the moft
infamous treachery, and the moft cruel
practices, are only looked upon as the
fineffes of a knowing enemy.

" This part of war, fays this great
" author, * has not been fufficiently
" explained; it were to be wifhed,
" however, that thefe kind of works
" were often read, and confidered by
" perfons of the profeffion. This read-
" ing appears to me to be the more ufe-
" ful; for, befides that it is amufing,
" it is alfo inftructive; and that when
" one knows the ftratagems and devices,
" he knows how either to render them
" ufelefs in the enemy, or to employ
" them himfelf when he has occafion."
To which I will add, that thofe opportunities often occur in the fpace of a
campaign; and that they offer every
day without being perceived, for want
of attention; nor are they known till after
they are paffed by. However, nothing
contributes more to the reputation of
an

* Notes on Polybius, tom. IV. p. 30.

an officer than thefe kind of actions;
but for that, fays Vegetius, you muft
lend a hand to fortune, and know how
to profit by the advantages fhe offers
you.

Among the furprizes of pofts that
may be executed by a fmall body, and
which an officer may undertake; there
are fome, to which we are invited by
their apparent facility, or by our watching the enemy with great attention.

I will not repeat on this fubject what
I faid before, of the precautions to be
taken in marching towards a poft; it
is enough to know that as thefe projects are like machines, where the dif-
placing a fingle wheel makes all the
others ufelefs; you muft examine with
the greateft nicety, every way and means
intended to be ufed, to fee that they
correfpond well with each other, in order to fucceed in the meafures you have
taken. How courageous foever the officers in our army may be, it muft not
be imagined that they are all capable to
embark in enterprizes of this kind; for
befides that they fhould be impene-
trably

trably fecret, and of great difcernment
in the choice of their foldiers; they
fhould alfo have a perfect knowledge of
the country, and be able to fpeak the
language; but few officers are poffeffed
of all thefe qualifications

As to the manner of furprizing a
poft, I have faid that it was impoffible
to eftablifh certain rules on this fubject, becaufe among a thoufand opportunities that chance offers to us, there
may not happen to be two alike. A
very quick march ftolen upon a diftant
poft, where the guard is negligent; a
thick fog which prevents them from
being feen; a river that has a ford un-
known to the enemy; an aqueduct, or
fubterraneous paffage, or a hollow way
that is not guarded; a little river frozen
over; a road ftopped up; good intelligence; the time of a fair; market-
day; and difguifes of all kinds. Such
are the different ftratagems which may
be ufed occafionally, and even promife
a happy iffue, though the fame have
been often made ufe of. I will only
obferve, that there are ftratagems where
it

it will be impoffible to fucceed, unlefs
you join therewith a fteady force. A.
confiderable poft, fuch as a town or
village, for example, where a party is
to be introduced by having an intelligence within, cannot be carried, unlefs one is well feconded. The only
means to conduct the furprize of thefe
pofts well, is to make it a rule to divide your force, to make yourfelf mafter
of the *chateau*, the church, church-yard,
and all the public places. It is a
miftake to fay that troops, divided in
this manner, could only act weakly,
and would run a rifk of being knocked
on the head one after another. I would
chufe always to make as many detachments as the enemy had pofts ; becaufe
in the fright, caufed by thefe furprizes,
it is eafy to make yourfelf mafter of thofe
pofts, before their defenders have time
to difpute them, or even to look about
them. The enemy being obliged to
divide their force alfo, and not knowing to what place to give the preference,
it is almoft a moral certainty, that being
ftunned with the noife on every fide,
C c they

they will even drop their arms out of
their hands : let us add alfo, that the
horrors of a dark night, and defpair
that never fails to poffefs a body of men
that are furprized, reprefenting objects
much greater than the reality, they will
imagine they have a whole army to encounter with.

The ill fuccefs of the furprize of Cremona in 1702, where the Germans had
divided their forces, proves nothing
againft my opinion. If without ftopping to make prifoners, they had
march'd a detachment directly to the
caftle, which ought always to be an object of principal attention in thofe kind
of actions, it would have been impoffible for the brave officers, who repulfed the Imperialifts, to have made
fo glorious a defence. M. de Schover,
who furprized Benevare in Spain, in
1708, took quite a different way, and
accordingly gained his point. This
General having learned that the Spaniards neglected the guard of an old
caftle, which was at the entrance of
that town, marched thither in the
night,

night, and made himself master of it, and afterwards sent sundry detachments to attack the Spaniards in the town; but these surprized by so sudden a visit, endeavouring to save themselves by flight, ran to the castle, as the last resource of a garrison; where, as fast as they entered, they were made prisoners. This method, to go strait to the castle of a town that one intends to surprize, is then the best to follow; because the enemy do not expect that the first attack will be made upon their strongest part; and being in equal concern both for the town and citadel, it is to be presumed that they will have divided their forces, so as to be able to defend all parts alike.

If any thing has been the cause that these events are not frequently seen now-a-days; it is, that they are almost always forgotten, and that the authors of them scarce ever obtain any recompence; nevertheless, what does not a man merit, who determines on an action of this kind, bravely adventuring himself to be sacrificed in the attempt?

M.

M. Menard has given us a relation of
a furprize,* in the hiftory of the town
of Nifmes, which deferves to be told,
for the good leffons it contains. Ni-
cholas Calviere, called Captain St. Cofme, having refolved to make himfelf
mafter of this town, agreed with a
miller, whofe mill was within the walls,
near one of the gates, called *de la Bou-
querie*, that he fhould file, during feveral
nights, an iron grate, that fhut up the
entrance of the aqueduct, thro' which
the fpring-water paffed into the town ;
that he fhould cover the parts cut by
the file with wax, fo that they might
not be perceived in the day time ; and
that he fhould conceal a hundred armed
men in his mill, while a more confiderable body of horfe and foot, that
were to arrive from Vivarais, advanced
to fupport the enterprize.

The orders for the rendezvous of the
troops being given, and the day for the
execution fixed to the 16th of November, 1569, St. Cofme fallied out of the
mill

* Hiftory of the town of Nifmes, tom 5, in
1569, and note 2d.

mill at three o'clock in the morning, advanced to the guard-houfe of *la Bou-querie*, killed all the foldiers, and opened the gate for two hundred horfemen, each of which carried a foot foldier behind him. Thefe men being got into Nifmes, immediately formed into feveral detachments, one whereof went to block up the caftle, whilft the others, going to all the fquares and open parts of the town, and founding their trumpets, became mafters of it in an inftant.

I thought this furprize of Nifmes the more neceffary to be related here, as its circumftances are very inftructive. Captain St. Cofme, who knew how to profit by the governor's neglecting to guard the entrance of an aqueduct; the choice he made of horfe to expedite the coming of his foot from different quarters; the exactnefs of his orders to thefe troops, which were about fifteen leagues from Nifmes, to be at the place of rendezvous at the hour appointed; his precaution in fending to inveft the caftle, to avoid having the garrifon to fight with in the ftreets; his attention

to

to diftribute his forces to all parts of the town, and to make them found trumpets, to make the inhabitants think their numbers were very great: all thefe circumftances are fo many inftructions for officers who may be tempted to undertake a like enterprize.

M. Carlet de la Roziere, engineer at the ifle of Bourbon, has collected fome examples of furprizes, which have fucceeded by a particular conduct. Brachio, a captain of Jean, queen of Naples, being defirous to become mafter of a tower in the territory of Ambrefa, difguifed one of his foldiers like a woman, giving him a bafket in his hand, together with a fickle. This man, fo transformed, ran as faft as he could towards the tower, feigning to fly for fear of fome of the enemy's parties; the guard let him in, and even let him go up a ladder to the top of the tower to fhew the centry where the enemy were. But as foon as he got up, he clove the centry's fkull with the fickle, and feized his arms, with which he drove off thofe that were below from their poft.

It

It is in this manner, that fometimes where it is impoffible to conquer by force, the mind fhould be attentive to profit by the leaft fault of the enemy. Epaminondas knowing that his wife was beloved by Phebiades, governor of Cadmia, ordered her to fup with the governor in the citadel, which he defended, and to invite a great many other ladies. The wife obeyed, and the guefts came to the rendezvous. But towards the end of the repaft, thofe ladies going out of the citadel to a nocturnal facrifice, which was only to laft a few minutes, the guards were ordered to let them pafs. As foon as they were out, Epaminondas ordered them to lend their cloaths to fome chofen foldiers, who being introduced to the citadel by one of thefe ladies who had the watch word, furprized the governor, and became mafters of the fortrefs.

Neceffity is in war, as every where elfe, the mother of invention, when one has firmnefs enough not to be difcouraged. The Amphictions befieged Cirrha; the greateft part of the inhabitants

bitants were fupplied with water from a
plentiful fpring, by means of an aqueduct. Eurilochus, * one of the Generals, having difcovered this aqueduct,
he ordered a great quantity of hellebore
to be mixed with the water; fcarcely
had the defenders of Cirrha drank of
it, when they were tormented with fuch
dreadful gripings, that they were incapable of defending themfelves; fo that
the Amphictions became mafters of the
place without the lofs of blood.

"The difproportion of forces," fays
M. le Chevalier Follard, fomewhere on
the fubject of furprizes, " is not always
" in the number, but often in the ca-
" pacity of the one, oppofed to the ig-
" norance or neglect of the other."
Mary, queen of England, not being
able to reduce the duke of Suffolk,
the head of the party that difputed
the

* Frontinus and Polyænus difagree on the fubject of the author of this device; Frontinus at-
tributes it to Colixthenes, alfo General of the
Amphictions, and Polyænus to Eurilochus.

the crown with her in 1553,* sent one hundred bold soldiers to the fortress where he was intrenched. These men reported themselves to him as deserters, that were coming to join his party; but as soon as they got in, they turned their arms against him, and seizing his person, gave him up to Mary, who beheaded him.

The civil wars of France, about the end of the sixteenth century, produced also some examples of posts being carried by stratagems: captain Martin, and du Rolet, governor of Pont-de-l'Arche, having formed a design on Louviers in Normandy, in 1591, surprized that town by means of a corporal, a priest, and a tradesman. The priest took upon him to keep watch in the belfry, † and to let the troops ad-

D d   vance

---

* This story rests entirely on the credit of some French writer, whence most probably M le Cointe has copied it; our histories, I believe, say nothing like it,

† It is always customary, especially in time of war, to keep a watchman, both day and night, on church steeples, to give notice of the appearance of troops, or of fire.

vance as near the town as they pleafed,
without ringing the alarm, and the two,
others promifed to deliver up the gate.
Thefe meafures being taken, Du Rolet fent forward feven refolute foldiers
with black fcarfs, which was always
worn by thofe of the league; thefe ftopped under the gate of the town, where
the corporal and the tradefman talked
with them, as with people of the union.
Du Rolet being informed by the tradefman that it was time to fall on, came
out of his ambufh, ran to the gate,
took poffeffion, cut the guard in pieces,
entered the town, and immediately became mafter of it, with the affiftance of
new troops brought to fuftain him by
the Baron de Biron.

There are none but thofe who have
the ftrongeft defire for glory, and whofe
valour is never diminifhed by danger,
who know how to reduce an enemy by
ftratagem, and to feize an opportunity
that fortune offers to them.

Guftavus Vafa, feeing the fea frozen
over, croffed it with his army, and went
in the middle of the night to burn the
naval

naval army of the Danes, that were ftopped at a little diftance from Stock-holm, whither they were going to increafe the power of tyrants, and the defpair of the people. *

I will add but one word more on the fubject of furprizes, on which one might write volumes; it is, that after one has formed a defign, and examined all its branches very well, one fhould never ftop in the middle of the execution, for having difcovered an obftacle that was not forefeen. Dionyfius having intelligence in the town of Naxos, appeared before the place at night with a confiderable body. The garrifon being informed of the treachery, took up arms and got on the ramparts; Dionyfius, though aftonifhed, was not difcouraged; on the contrary, he threatened to put all to the fword; and fent a boat into the harbour with a certain number of

D d 2   boat-

---

* Amazing! that fuch an expreffion could fall from the pen of a French foldier, when he and all his companions have been fighting for a century paft, only to fwell the vanity and importance of the tyrants that enflave them.

boatfwains, (or thofe who keep the gal-
ley rowers to their work,) with their
whiftles; and as each gave his fignal
different, the Naxiens imagined that
there were as many galleys in the port
as they heard whiftles, and furrendered
at difcretion. If Dionyfius had endeavoured to retreat on feeing his defign
difcovered, he would have been greatly
expofed : for befides, that the Naxiens
could have come out, and cut off his
rear guard, he would have been alfo the
object of their railleries. After the battle of Cannæ, Hannibal advanced as far
as the gates of Rome, with a defign to
lay fiege to it. But he was hindered by
a great noife he heard in the night, like
a number of people laughing very hear-
tily : the Romans being aftonifhed the
day after at his retreat, built a temple
immediately, which they dedicated (Deo
ridiculo) to the god of laughter.

I will fay no more on the feizure of
pofts by ftratagems, becaufe one may
fee by what I have faid, that thefe actions in general are not fo difficult as
they are commonly thought to be. Ti-
morous

[205]

morous people, and thofe who in the
fmalleft affairs are ftopped by the moft
trifling difficulty, may probably look
upon them as impoffible, and even think
there is fomething fupernatural in thofe
they fee fucceed ; but it is not for fuch
people I write. To men of application,
of bravery, and bright genius ; and in
a word, to the officers of our own times,
who are now in fervice ; to them I fubmit my ideas, and them I have chofen
to judge of the methods I have herein
propofed.

The end of M. le Cointe's treatife.

Monfieur

Monſieur de la Croix, who publiſhed a
ſmall treatiſe in 1752, intitled, *Traité
de là petite guerre*, or a treatiſe on
petty war, gives the following account of his deſign.

" THE different actions in which
I have been engaged, during
fifty years that I have had the honour
of ſerving the king; the important expeditions of which I ſhared in the execution with my late father, *(a major
general)* have given me experience in
the *petite guerre :* I have endeavoured
to learn every thing that could contribute to raiſe the ſervice of *free companies* * to the greateſt degree of perfec-
tion : I have laboured to remove inconveniencies, to reform abuſes, and to
eſtabliſh an order and diſcipline ſuitable
and

* A *free company* is one of thoſe corps commonly called *irregular*, is ſeldom or never under
the ſame orders with the regular corps of the army,
but for the moſt part acts like a detached army,
either by itſelf, or in conjunction with ſome of
its own kind ; therefore their operations are properly conſidered under the title of the petty war.

and proper for that *corps* ; I have taken pains to gain the confidence both of officer and soldier ; to know their qualities and their virtues, in order to employ them usefully. Succeſs in divers encounters has taught me how to attack, and how to avail myſelf of ſtratagems of war on all occaſions that occur, and in all conjunctures and countries that I happen to be in. I thought I could not make a better uſe of my little knowledge than to publiſh it, in order that thoſe who enter into the military profeſſion may thereby improve themſelves to the advantage of the ſervice."

But as a great part of M. de la Croix's treatiſe ſeems to be a pious panegyric on the actions of his father, the tranſlator propoſes only to extract a few paſſages, and ſuch as may conſonantly be added to the foregoing treatiſe.

De la Croix's precautions for a march.

Stratagems are to be employed in all enterprizes againſt the enemy ; but it is is also essential to dive into those that they may use against you, so that you may be able to elude and counteract their effects.

When any important and distant expeditions are in agitation, and when it is required to send great detachments so far as sixty leagues or more, you should begin by taking the best instructions you can, relative to the country through which you are to pass, what routes you are to take, the situation of the enemy, of the strength and situation of the different posts possessed by them; and on your departure from your garrison, you are to see that every man comes out with every thing necessary for the expedition; (but in order to deceive the enemies spies,) you may form many small detachments of twenty-five, thirty, or forty men, to the number of two or three hundred, which may march out by different gates, and on different days.

These divisions are each to be headed by an officer, who being informed of the march of the other detachments,

must

muſt regulate the days of his march, ſo
that they may all meet nearly at the
ſame time at the place appointed for
their re-union, by a certain ſignal
agreed on. Here a general inſpection
is to be made of the whole body, to
ſee that none have deſerted; after which
they are to march, by filing off in great
ſilence, and keeping the by-roads; the
villages ſhould be avoided, and they
ſhould only march at night, and not ex-
ceed four leagues at moſt; at day-
break they ſhould throw themſelves in
ambuſh, into a wood, to wait till night,
and refreſh themſelves with the provi-
ſions that they took care to furniſh
themſelves with before their march.
But theſe precautions, however wiſe
and prudent they may be, are not near
ſufficient for the ſecurity of a body that
are to penetrate into an enemy's
country; the commanding officer ought
not only to be careful to make the
march ſilent and ſecret, but he ſhould
foreſee how he is to return: he ſhould
never move one ſtep till he has pro-
jected the means of his retreat: the ex-

E e   ecution

ecution of his defign ought to engage
him to keep up a moſt difcreet and mo-
deſt behaviour towards the country
people wherever he is to paſs, and to
make the march as little burthenſome
to them as poſſible ; he muſt gain their
eſteem by his affability, good order,
and the ſobriety of his men. This muſt
be followed by ſome preſents prudently
beſtowed, which will not only prevent
his being harraſſed on his march, but
will diſpoſe the country people to diſ-
cover ſome important particulars to him,
and ſhew him the motions of the enemy,
their numbers, poſition and ſtrength.
Theſe inſtructions being confirmed by
ſpies, and the people that he has ſent
before to examine the country, will enable him to act with ſome confidence,
and make him almoſt ſure of the ſucceſs of all his enterprizes.

To prevent detachments from being
diſcovered by the barking of dogs,
when they are paſſing by farm-houſes
that lie ſeparated from villages, the officer ſhould order ſome men diſguiſed
to go before, furniſhed with poiſoned
pills,

pills, or *nux vomica*, for thofe animals.
Thefe people ferve as well as fpies to
make difcoveries of confequence, for the
fecurity of a march.

*The ufe of infantry, and the utility of
horfe in a retreat.*

All confiderable expeditions are made
by the foot. The horfe intended for
this fervice are not to fet out till fome
time after the foot, they are to approach near the place where the blow
is to be ftruck, to lie in ambufh till the
moment of its execution; and then, for
the moft part, have no more to do than
to fuftain the foot in their retreat, which
is to be made by long marches; and in
order to make this retreat more fecure
and eafy, they fhould endeavour to procure, in the places they pafs through, a
fufficient number of good waggons and
ftout teams of horfes, which helping
the foot to forward their march more
brifkly, may alfo ferve as a barricade
againft the enemy's horfe that might
purfue them into the plains.

E e 2   *M.*

M. de la Croix alluding to his father's and his own past experience, shews how the most difficult things used to be executed.

The moft difficult projects were formed and were executed; difficulties never difcouraged us, all obftacles were furmounted, and the enterprize had a happy iffue. The reafon is plain, the troops were experienced and accuftomed to war; the officers had good under-ftanding, and were men of honour; the commanders in chief were affured of the merit of thofe they employed; their method alfo was not to be rafh, know-ing that immoderate heat and too much precipitation, inftead of advancing the fuccefs, would make it mifcarry; and that by trufting too much to chance, the beft enterprizes may fail. They knew how to temporife wifely, to allow their project time to ripen. They had trufty people in different places, whom they rewarded with great punctuality, who furnifhed them with exact accounts. Their leaft ftep and all their motions were

were guided by prudence; their operations had always a fuccefsful end. What better maxims can be laid down in the military art?

Care and precautions to be taken in towns, villages, and places of refreshment.

A body or detachment are conducted as their chief thinks proper, and put an implicit confidence in him, when he has given them proofs of his vigilance and attention to their fafety. Therefore when he enters a village or town to refrefh, he fhould immediately poft double centries in the fteeples, or higheft buildings, which are proper to make difcoveries from, to obferve the environs, to prevent furprizes and unforefeen attacks; then he is to diftribute the provifions, and give out the neceffary orders to his people; he is not to confine himfelf to this alone, he muft artfully pick up ufeful and neceffary intelligence; he muft talk with the burgomafter and other principal people of the place, endeavouring by obliging

means

means to gain their confidence, to draw
from them fome interefting confeffions;
he is to demand of them trufty perfons
to fend before him, and to promife to
pay well for any fervices that they may
do for him: laftly, he muft fpare neither money or pains; the money is moft
efficacious, it muft be liberally difpofed
of on proper occafions, and without re-
gret; and the returns will be ample in
the advantages that will refult from it.

*Other precautions and meafures for night
marches; attention to the fire arms;
and the effential cuftom for retreats.*

I have faid before that night is the
beft time for a march; and it cannot
be fecret at any other time; but great
care muft be taken in the dark. A body
fhould file off flowly, regularly, and in
filence; the commander fhould order
halts from time to time for indifpenfa-
ble neceffities, and order the officers to
watch while the men are marching in a
file, left they miftake one road for another, and to remain in the rear till fuch
time

time as they are again rejoined in a body or column. No man muft be fuffered to fmoak, even in the ambufcades, on account of the inconveniencies of the fmoak and the fmell of the tobacco : if they are to pafs through plowed ground they fhould drag large faggots of briars after them to efface the marks of their feet, left the peafants fhould obferve them : when they arrive in a wood at day-break, as the leaves are commonly covered with dew or rain, thofe at the head fhould cárry a kind of blinds of oiled cloath to cover them, to break way for the reft to follow, fo that they fhall not be wet.

To keep the fire arms in good order, and to prevent them from being wet in rainy weather, the method is to have a fmall cafe to draw over the butt to cover the lock of the piece; this is foon pulled on and off, and will keep the lock and priming in order to fire, which is not always the cafe ; for how many regiments when it rains, march without attention, and out of order, the foldiers carrying their butts behind expofed to the rain, and

may

may be attacked by a much lefs body ?
examples of which have been feen.
This caution is not all ; the ferjeants
fhould vifit the men's arms every day ;
and as ammunition is in one refpect
more precious than provifions, it fhould
be managed with the greateft care. A
foldier ought to have at leaft a hundred
rounds to fire ; and he fhould never dif-
charge one without effect, as *free companies* have no train waggons to attend
them.

Laftly, as we ought to forefee every
circumftance, we fhould be provided
againft every accident ; we fhould carry
grenades, combuftible ftuff, caltrops,
nails to fpike up cannon, petards, hatchets, fhovels, match, and jointed har-
rows : the ufe of thefe things is foon
known, they ferve to burn forage, and
hinder and delay the purfuit of cavalry
in a retreat.

*An ufeful maxim for rencounters, nocturnal
and unforefeen attacks.*

All the devices and precautions ufed
by the commander of a body, are intended

tended to perfect the project he has formed; this same reason should make him attentive to attempt nothing on the enemy that may thwart or delay the execution of that project. Sometimes an opportunity offers to fight, and even crush a detachment passing by his ambuscade; but he must be careful not to take this advantage, lest he thereby obstruct his project; and if by accident he should meet a body at night, he should proceed thus, in order to gain his point. His advance guard should be preceded by two or three men, who should go on very silently, and stop now and then to listen; if they hear any thing, they should come without noise to give an account thereof; but if they should unexpectedly fall in with the enemy, they should call out loudly *who comes there?* At this noise the body fix their bayonets, keep close, and throw themselves to one side, on the right or left of the road, to wait the result. The commander attentive to the motions of the enemy, who, as it often happens, might have only come to-

F f   wards

wards him by chance, lets them pass by;
but if they come on imprudently, tho'
they may be superior in numbers, he is
in a condition to receive them with his
body, whom he makes kneel with fixed
bayonets. However, these are accidents
which it is always prudent to avoid;
for though one may come off victorious
from such an encounter, yet the body
is weakened thereby, and these kind of
successes often deprive you of the power
of executing the only project proposed.

A stratagem commonly made use of by M.
de la Croix.

The orders being given to set out a
detachment of three or four hundred
men, they marched by divisions, as be-
forementioned, of thirty, forty, or sixty
men, and by different roads, but to
meet on a fixed day; then they kept
close in woods, or where they could lie
hid; and a body of sixty or eighty men
were detached to go and refresh themselves in the neighbourhood of the place
where they wanted to allure the enemy:
the

the commander or governor of the
nearest town, on hearing of this body's
approach, from the inhabitants of the
place, never failed to send out a large
detachment, in proportion to the numbers he heard were in the country; many
have been very roughly treated who did
not expect to be attacked by more than
the reports gave out, till they were
drawn into the ambuscade by the allur-
ing party.

The advantage of night attacks, and the
precautions to be taken in quarters.

Night attacks are almost always successful, and the reason is pretty evident.
The assailants are informed of the position and the strength of the enemy; the
latter are ignorant both of the numbers
and of the manœuvres that are to be
employed against them; the one knows
where to strike, and is sure of his blow;
the other hardly knows what part he is
to defend: in these circumstances whole
battalions have been beat and routed
by moderate detachments. There are
some,

some, who confiding in their numbers,
and the valour of their men, and satis-
fied to be told that there is no considerable body of the enemy near them,
abandon themselves to their ease, and
cannot be persuaded that two or three
hundred men could come to insult
them; in this false opinion, as soon as
they arrive at a town or village, the
commander, after having ordered the
quarters, appointed the posts, and placed
the guards, seeks a good lodging for
himself, and gives himself up to his
ease; the other officers follow his example, and take care to want for nothing
comfortable; and all indulge effemi-
nately in the middle of danger: but
they often pay very dear for such im-
prudent conduct; the enemy, who are
on the watch, are informed of their arrival; spies bring them news of the true
state of things every where, and they
soon become acquainted with the position of the advanced guards, and of the
commander's quarters.

These kind of enterprizes have always been looked upon as very bold
and

and even rafh, to dare to attack a body
of fix or feven hundred men, with a
detachment of two or three hundred;
yet it is not to be doubted but a true
partifan, who is well acquainted with
the country, and with the march of a
fuperior body, may eafily form his attack in the dead of the night, and better in bad weather than in good, as he
has his arms always dry by the method
I mentioned before, let the weather be
as it may; on fuch an occafion he arrives at a village, with his party at the
diftance of a league or more from the
enemy, where, during his halt, he informs
himfelf, by the chief magiftrate, of every
particular, who will not difobey him;
he is alfo to afk for fome men of the
place to ferve to help him to reconnoiter
the enemy; fuch people are always to
be found, who for a proper recompence, or from an inclination to be contrary to different troops, are eafily determined to this fervice; they are to be
inftructed what to do, and what to obferve; to know where the guards are
pofted; where the commander is lodged;

if

if there are no ways of furprizing them
by going behind through fome gardens;
he fhould afk them if they have got any
relations in the places to name, in cafe
the guards fhould ftop them, and fo
take off all fufpicion. After thefe meafures, they are to be ordered to return
to an appointed place, when they are
ready to make their report. Thofe expeditions feldom fail; and to fucceed,
the body is to be divided into three or
four detachments, with a view to fall
on all at once, and not to give the enemy
time to look about them; but fhould
any one fay, what confufion at night?
How can thefe detachments join again?
The anfwer is, that tuly thefe attacks
are very hazardous to both one fide and
the other; but the affailants are never
embarraffed for the following reafon,
which is eafy to be conceived, that before the attack they take care to fend
eight or ten foldiers, each carrying a
trufs or two of ftraw on a ftake, to fet
fire to at the moment of the attack;
this fire ferves as a direction to thofe who
attack to retire to the light after they
                                    have

have taken some prisoners. All these kind of attacks are made in less than half an hour, and the enemy cannot know the meaning of the fires; and this device hinders them from observing those who attack them, so as to be able to pursue them in their retreat.

A commanding officer cannot be too circumspect in and about his quarters, especially while ever he is an enemy's country, where the natural aversion of the inhabitants will be joined to the activity of the enemy to harrass and overpower him.

The secret of marching small divided bodies of a detachment, and to make them rejoin quickly at the appointed time and place, is of infinite advantage, and puts them in a condition to form their attack with more certainty of success, as the enemy don't expect to have to do with a large body; they are not concerned when they hear of forty or fifty men only in the field: if they are even told of another body of the same number, that have been seen, they are persuaded that it is the same they heard
of

of before; and they are seldom unde-
ceived, till the time that the union of
the whole is made, and ready to begin
an attack, which they never apprehended.

End of the extracts from M. de la Croix.

Some

Plate 1ˢᵗ

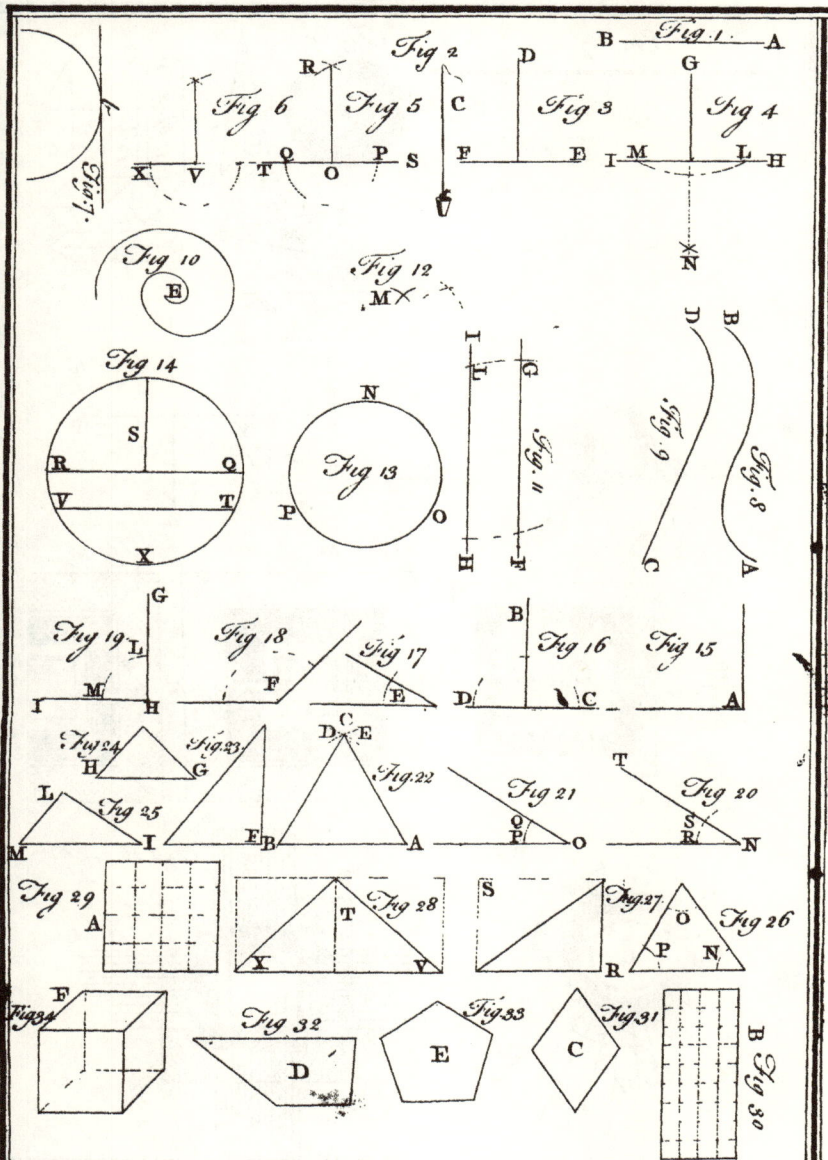

Fig 1.
Fig 2
Fig 5
Fig 6
Fig 3
Fig 4
Fig 7
Fig 10
Fig 12
Fig 14
Fig 13
Fig 11
Fig 9
Fig 8
Fig 19
Fig 18
Fig 17
Fig 16
Fig 15
Fig 24
Fig 23
Fig 22
Fig 21
Fig 20
Fig 25
Fig 29
Fig 28
Fig 27
Fig 26
Fig 34
Fig 32
Fig 33
Fig 31
Fig 30

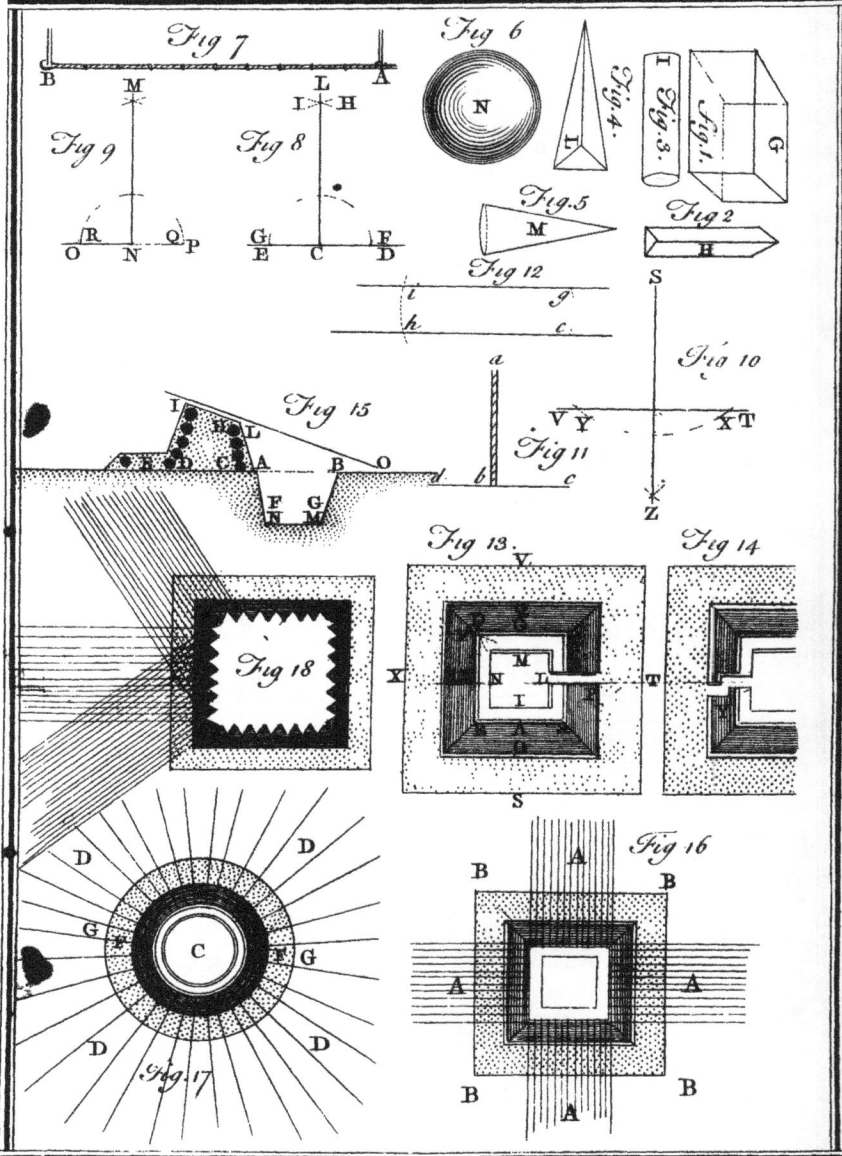

Plate 2ᵈ

Plate 3 d

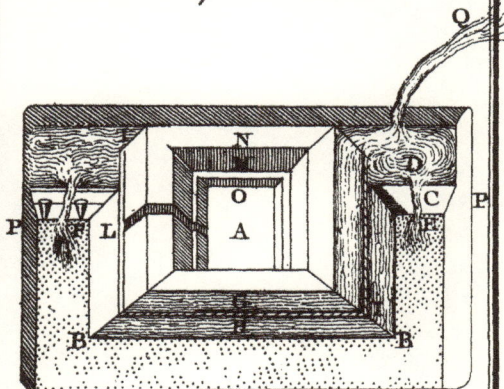

The View of a Redoubt

A The inner Ground of the Redoubt
B The bottom of the Ditch
C.D.E The damm of earth
F A dam of boards planks or fascines
G { The upper part of the Redoubt made with fascines
   { or with earth thrown out of the ditch
H The lower part of the Redoubt cut into the earth
I { The berme or space, left at the outer bottom
   { of the Parapet to keep up the Earth
L The entrance of the Redoubt
M The inside of the Parapet
N The outside of the Parapet
O The Banquette
P The Glacis
Q The river introduc'd to fill the ditch with water

Some hints and observations borrowed from Marſhal Saxe.*

Of war among mountains.

THERE are few things to be ſaid on this ſubject. But thoſe who make war in mountains ſhould be extreamly cautious; they ſhould never venture into the valleys without firſt poſſeſſing the heights, then all ambuſcades are at an end, and they may paſs

G g on

* The tranſlator is ſenſible that the foregoing extracts, as well as theſe hints borrowed from M. Saxe, relating more to the duty of a *general* than of a *private* officer, riſe above the plan laid down by Monſ. le Cointe, and may ſeem improperly added to ſuch a work; but as in the Britiſh ſervice every ſubaltern ſhould expect to riſe to the higheſt military rank, he thought it not amiſs to open the proſpect at the end of M. le Cointe, and to give them a ſmall view of the path they are to purſue in their ſtudies for the higher offices, whereby they may ſee that the duty of a *general* officer is only the exerciſe of the *machine* at large, whereof that of the *ſubaltern* is the *model* in miniature.

on in safety; without this precaution there are great rifques of being knocked on the head, or of being obliged to return, after lofing a great many men.

But if you find the paffages, as well as the heights pre-occupied, you fhould make a feint to force them, to amufe and attract the attention of the enemy, and at the fame time feek a pafs through fome other way. Though difficult the mountains may appear, one may always by ftrict fearch find paffages. The people who inhabit them do not know them themfelves, becaufe they have not been obliged to feek them; therefore one fhould not believe their reports, who, for the moft part, are acquainted with the things relating to their country by tradition only : I have often found out their ignorance, and the falfity of their relations.

In fuch a cafe one fhould fee and fearch one's felf, or employ people who will not ftop at difficulties; thefe things are always found when diligently looked for; and the enemy, ignorant thereof, runs away, not knowing what courfe to
take,

take, becaufe he only provides for common things, that is, againft the moft
practicable roads.

Of a country inclofed by hedges and ditches.

As the enemy is as much embarraffed
in this kind of country as you can be,
there is not much to be apprehended;
thefe are little matters that decide nothing; and where the moft obftinate
will fucceed. There is only one thing
to be minded; it is to keep all behind
clear, fo as to be able to detach, and
to retire, in cafe of neceffity. Here it
is very neceffary to know how to place
cannon, which is of great fervice : as
the enemy dare not ftir from the pofts
they poffefs, one may cannonade them
with eafe; if they abandon them, their
retreat is not always fuccefsful, and one
has fometimes the good luck to cut
them off.

But as I faid before, that thofe things
in the whole are not very decifive, they
fhould be guided by the fituation of the
places; fo that no certain rules can be

pre-

preſcribed. Neverthelefs one ſhould always obſerve this, to puſh forward, and on your flanks in marches, with detachments of one hundred men, ſupported by double, and the double by a triple, ſo as to be covered and in ſafety.

A detachment of ſix hundred men will ſtop an army; becauſe on the highways that are incloſed by hedges and ditches, ſuch as are found in Italy, and in all well watered countries, one ſhews a great front to the enemy, who will ſuppoſe your numbers to be much greater than they are. The leaſt hut makes a fortification where one may ſupport a very obſtinate engagement, which will give one time to look about, and to make a diſpoſition; for we ſhould guard againſt ſurprizes in theſe kind of countries.

A partiſan of ſpirit and addreſs, with three or four hundred men, will cauſe a frightful diſorder, and will attack an army very well on its march: If he cuts off the baggage at the beginning of the night, he will carry off a great part of it, without running much riſque; becauſe

cauſe if he retires between two ditches, and ſecures his rear, by blocking up or otherwiſe embarraſſing the road, he will ſtop you ; if he is puſhed, he leads the waggons in a line, and the firſt houſe he finds, he ſtops you ſhort, during which time the baggage he has taken advances in the country. If he acts thus againſt your horſe, he will put them into great diſorder ; it is for this reaſon that you ſhould have advanced, and flanking parties in and about all the avenues of your march, and they ſhould not be weak ; for there is no doubt but the enemy will be lying in wait for you on ſuch an occaſion ; and you muſt fight to the laſt minute to avoid diſhonourable ſurprizes. If you have to do with an enemy, whoſe general has common ſenſe, he will ſoon find people in his army, who have bold and penetrating minds, ſo as to ſee things juſt as they are.

Of

Of the paffage of rivers.

It is not fo eafy as one imagines to hinder the enemy to pafs a river; but he can come to attack you much eafier, than he can defend himfelf when he is retiring before you. In one of thofe cafes he fhews you his front, and fupports himfelf by a good difpofition, and by a hot cannonade; but in his retreat he fhews you his rear, which is very difficult to defend, and much more fo as he is hurried; and as this difpofition is never fo well made as that of an attack; and as every body in a retreat contracts a kind of timidity, which makes them feel already half vanquifhed.

In regard to the paffage of a river by main force, I believe it is not poffible to prevent it, efpecially when it is fupported by a brifk cannonade, which gives time to the head to intrench themfelves, and to make a work to cover the bridge. There is nothing to be done

done in the day-time, but at night one might attack this work; and if it fhould happen to be at the time that the ene-my's army begins to pafs over, confu-fion fpreads every where, and thofe who have got over already will be loft; but one fhould attack with ftrength; and if you let the night flip, you'll find all the enemy paffed over the next day. Then it is no fmall action, but a general one, that is to be rifked, and which reafons of ftate do not always permit.

F I N I S.

www.ingramcontent.com/pod-product-compliance
Lightning Source LLC
Chambersburg PA
CBHW030406100426
42812CB00028B/2846/J